PUPPET MASTERS OF EVOLUTION

CONTENTS

PUPPET MASTERS OF EVOLUTION

HOW PARASITES SHAPE LIFE

ANDREW A. DAVINACK

Columbia University Press *New York*

Columbia University Press
Publishers Since 1893
New York   Chichester, West Sussex
cup.columbia.edu

Cataloging-in-Publication Data is available from the Library of Congress.

ISBN 9780231223263 (hardback)
ISBN 9780231223270 (trade paperback)
ISBN 9780231564700 (epub)
ISBN 9780231567725 (PDF)

LCCN 2026029936

Printed and bound by CPI Group (UK) Ltd, Croydon, CR0 4YY

Cover design: Henry Sene Yee
Cover image: Shutterstock

GPSR Authorized Representative: Easy Access System Europe,
Mustamäe tee 50, 10621 Tallinn, Estonia, gpsr.requests@easproject.com

To my parents, Anthony and Jasoda David, who never fully understood but always supported my strange interests and hobbies;

to my wife, Dr. Jennifer Davinack, for her unwavering support throughout this project;

and to my late childhood friend, Reuel Lau, who left too soon: non omnis moriar.

PREFACE

I first encountered parasites seriously as a graduate student at Hofstra University, in a parasitology course. Alongside the scientific texts, our professor assigned Carl Zimmer's *Parasite Rex.* I remember being struck by how vividly Zimmer captured these hidden creatures—not as footnotes to biology, but as protagonists with the power to terrify, astonish, and captivate. His book helped spark a wave of public fascination with parasites at the turn of the millennium. For the first time, I saw parasitology not just as a medical specialty, but as a lens through which to view the entire story of life. That book planted the seed.

Years later, I found myself not just working with parasites but devoting my career to them. I became a marine parasitologist, specializing in worms that burrow into oysters and mussels, creatures so small and unassuming they are invisible to most eyes but can be devastating to fisheries and aquaculture. I learned quickly that parasites are everywhere—not an oddity of tropical medicine or a footnote in ecology, but a defining feature of biology itself. To study them is to study the hidden architecture of ecosystems.

The first inkling that I might one day write a book on this subject did not come in a laboratory or while poring over papers in a library. It came one winter night in upstate New York, during my second

which describes the relentless evolutionary arms race between hosts and parasites, reappear in multiple chapters because they are so central to understanding parasitism in an evolutionary context. For those intrepid readers who would like to dive deeper into the actual research, an extensive notes section has been curated at the end, providing details on each discovery and experimental result reported throughout the book. What unites all these strands is a conviction that parasites deserve more respect—not as benevolent creatures (they are not), but as forces of evolutionary innovation. They are sculptors of genomes, shapers of ecosystems, even hidden coauthors of human history. To tell their story is to tell *our* story.

A final note about the moment in which this book was written. As I draft these pages, the future of basic research in the United States feels uncertain. Proposals have surfaced to cut the budget of the National Science Foundation—the backbone of fundamental discovery—by half. That would mean fewer funded projects, fewer graduate fellowships, and fewer opportunities for the next generation of scientists who will map parasite diversity, decode ancient epidemics, and build the tools we need for the challenges ahead. Almost all of the work you will read about in these chapters was made possible by public investment in basic science: grants that did not promise immediate payoff but produced the knowledge that later saved crops, protected fisheries, guided vaccine and antibiotic development, and shaped data-driven public health policy.

Pure science is a long game. Today's obscure dataset becomes tomorrow's breakthrough. When we shortchange basic research, we do not just slow progress; we forfeit discoveries we cannot yet name. I have already watched talented students and colleagues look abroad for the support they cannot find at home here in the United States. If there is a plea tucked into this preface, it is simply this: invest in curiosity. Fund the careful, open-ended work that lets us ask better questions about the living world, including its strangest inhabitants.

This book is my attempt to return to that crowded cafe and keep the conversation going, to take gasps and groans and laughter and turn them into understanding. Parasites will still horrify you in these pages, but I hope they will also surprise you, provoke you, and ultimately deepen your sense of how life works. By the end, you may find that the organisms we have spent centuries trying to eradicate have been among evolution's most creative engineers. And if you come away with a little more respect for them—and a lot more curiosity—then that winter night's suggestion from a firefighter in upstate New York will have done its work.

Welcome to the hidden majority.

PUPPET MASTERS OF EVOLUTION

INTRODUCTION

The history of life on Earth is a history of parasites.

—Carl Zimmer

Autumn had finally come to New England. From the large window next to my lectern on the third floor of the Discovery Center, I watched golden leaves swirl in the crisp air as students bustled across campus. Some were hurrying from their Greek classes, others heading toward organic chemistry in the classroom next to mine. It was the fall semester of 2022—my first as a faculty member in the biology department at Wheaton College, a small liberal arts institution nestled in a semirural enclave of southern Massachusetts, about an hour and a half south of Boston and just twenty minutes north of Providence, Rhode Island. Students were returning from fall recess, and the campus was once again alive with chatter: talk of looming final exams, Halloween parties in Providence, and the countdown to Thanksgiving break. But as my students trickled into the classroom that day, I knew it was going to be one of *those* teaching days. Professors recognize the signs: weary faces, sluggish movements, a collective resistance to engagement. On days like this, a fifty-minute lecture can stretch into an eternity.

This is when instructors reach for their bag of tricks: clever pedagogical strategies plucked from teaching workshops, the latest buzzwords from academic literature, or perhaps a hastily implemented discussion exercise to force participation. I, however, had no need for smoke and mirrors. I do not pretend to be Bill Nye or Neil deGrasse Tyson, and discussions about "backward design," "assessments," or—God forbid—rubrics, bore me just as much as they bore my students. No, the ace up my sleeve was the course itself. That semester, I was teaching a course called Ecology of Infectious Diseases, a class that, by its very nature, required no gimmicks. It was a course about parasites, the diseases they cause, and the ecological systems that facilitate their spread. My graduate adviser, Jason Williams, at Hofstra University once said of parasitology courses: "They sell themselves." And so, instead of forcing discussion, I simply pressed Play.

"Alright, folks, welcome back. Today's going to be a reflection class—we're going to have a group discussion."

Predictably, groans rippled through the room.

Then I hit *Play* on a twelve-minute video clip.

On the screen, five surgeons in full surgical gear stood around an anesthetized patient. One surgeon held a metal bedpan, positioning it just below a partially opened section of the patient's abdomen. Another surgeon, clad in latex gloves, reached into the lower stomach and pulled out a writhing mass of peach-colored worms. The moment he dropped them into the pan, an audible *slop* filled the operating room.

The classroom erupted.

"Oh my god!" one student gasped, recoiling.

"What the *expletive*?" another blurted, wearing a strange mix of horror, fascination, and excitement.

But it got better.

The surgeon reached for another bedpan as his colleague removed yet another massive handful of worms. The creatures squirmed and twisted violently; their soft bodies still very much alive.

"I am NEVER eating ramen again!" someone shouted at the very back of the classroom.

Laughter ensued.

I scanned the classroom. The transformation was instant—every student was now fully locked in.

I paused the video.

"So," I said, letting the tension settle. "This patient came in complaining of persistent bowel pain. The culprit?" I clicked to the next slide—the CDC's (Center for Disease Control) cartoon-ish life cycle diagram of the intestinal roundworm, species name: *Ascaris lumbricoides*.

It is the largest multicellular parasite to infect humans. The eggs—microscopic and nearly indestructible—are expelled into the environment via human feces. Ingestion of these eggs leads to infection, and in places with poor sanitation, reinfection can happen over and over again.

A student raised her hand.

"Professor, how is it even possible for someone to have that many worms inside them? Like, do they just keep getting reinfected?"

Another student quickly pulled up a paper on Google Scholar on her phone and raised her hand.

"The eggs are super resistant. They can survive dehydration, acidic environments—even sewage sludge. If someone lives in unsanitary conditions, they can swallow them repeatedly."

Another hand shot up.

"Wait—do the worms stay in the stomach?"

Before I could answer, a student flipped open her parasitology textbook and held up an illustration. It depicted a sleeping person with a giant roundworm half-emerging from their ear.

Cue more exclamations of disgust.

Indeed, intestinal roundworms are capable of wandering, sometimes exiting through the nose, mouth, or rectum—causing far more psychological trauma than physical harm.

As the discussion escalated, students eagerly debated the roundworm's life cycle, host adaptations, and transmission dynamics. I barely had to say a word. I simply stood back, nodding occasionally, reveling in that rare and beautiful symphony that every educator longs to hear—self-directed learning.

Parasites evoke a visceral reaction in all of us. They disgust us, horrify us—and yet we cannot look away. They have inspired myths, shaped cultures, and even become weaponized metaphors in political discourse. But even the word *parasite* has a strange and ironic history. It comes from the Greek *parasitos*, meaning "one who eats at another's table"—*para* (beside) + *sitos* (grain or food). Originally, it referred not to something repulsive or grotesque, but to temple officials or civic guests who dined alongside priests during religious ceremonies. Over time, however, the term took on a more cynical tone in Roman and medieval societies, coming to describe social freeloaders, those who benefited from the generosity of others without giving anything in return. It wasn't until the eighteenth century that the word took on its modern biological meaning: an organism that lives in or on another, at their expense. That shift, from ritual companion to evolutionary villain, reflects our own uneasy relationship with parasites. We revile them not just because they live within and feed on others, but because they violate the moral boundary that we like to believe separates cooperation from exploitation. The fascination they elicit transcends expertise: the same shock, curiosity, and unease I see in junior and senior biology majors—students who may one day become scientists or medical professionals—are mirrored in the reactions of lay audiences when I give public talks on this subject. Look no further than popular TV documentaries, movies, and books. Animal Planet's *Monsters Inside of Me*, a documentary series that dramatized real case studies of parasitic infections, had a steady dedicated viewership from 2009 to 2017. Popular science fiction movies such as *Invasion of the Body Snatchers*, *Alien*, *Species*, *The*

students are often told, outside of traditional food webs. But this is a fundamental misunderstanding of their role. Rather than being outsiders, parasites are essential regulators of ecosystems. They control host populations, drive biodiversity, and maintain balance within trophic networks. In some cases, they even dictate the rise and fall of predator-prey relationships more than apex predators do. Without parasites, many ecosystems would collapse into chaos, overrun by unchecked species that would otherwise be kept in balance.[4] In short, parasites are not just passengers in evolution, they are its architects, quietly shaping the biological world in ways we are only beginning to understand.

Over the next few chapters, we will explore the profound impact parasites have had on evolution, biodiversity, and human civilization. Through case studies, cutting-edge research, and the bizarre but fascinating world of parasitic life, I will make the case that parasites are not mere villains, but rather architects of life itself. In chapter 1, I begin by examining how parasites sculpt the physical and behavioral traits of their hosts across evolutionary timescales—from parasitic barnacles that can feminize male crabs, turning them into reproductive puppets, to immature tapeworms becoming cancerous in an unsuspecting host. These examples reveal a hidden world in which parasites do not just take advantage of their hosts, they fundamentally rewire them, pushing the boundaries of what is possible in evolution. In chapter 2, I turn to the hidden genetic battleground, exploring how parasites have shaped our genomes over evolutionary timescales. We trace the scars they left in our DNA—the molecular remnants of ancient wars fought between parasites and our ancestors, conflicts that ultimately forged one of the most intricate biological systems on Earth: the immune system. In chapter 3, I uncover the hidden role parasites have played in the evolution of complex life on Earth. Many parasites engage in horizontal gene transfer, a process whereby genetic material jumps between species, bypassing traditional inheritance. This chapter will reveal how parasites may have accelerated the development of complex multicellular life and contributed to antibiotic resistance. In chapter 4, I examine the mind-bending ways parasites control their hosts. From fungi that turn insects into zombies to parasitic wasps that turn caterpillars into bodyguards for their offspring, parasites have evolved astonishingly precise ways to manipulate their hosts' behavior, physiology, and gene expression.

But what about humans? Emerging research suggests that some parasites may subtly influence human behavior, raising deeper questions about free will and evolutionary pressures on our own species. In chapter 5, I investigate the evolutionary battle between parasites and their hosts. Just as parasites develop ever more elaborate ways to evade immune systems, hosts evolve countermeasures to resist them. This ongoing conflict has shaped everything from immune system complexity to reproductive strategies and social behaviors. In chapter 6, I challenge the common misconception that parasites are unchanging agents of destruction. In reality, the boundaries among parasitism, mutualism, and commensalism are far more fluid than we often assume. Parasitism exists on a spectrum: what begins as parasitic exploitation can evolve into cooperation. This chapter is particularly important as we will explore how mutualistic relationships could become parasitic and how even harmful parasites may have unexpected benefits for their hosts. Could parasites actually be essential for life as we know it? In chapter 7, I examine how parasites have shaped not only biological evolution but also the course of human civilization. From malaria influencing the outcome of wars to parasites affecting human migration patterns and cultural taboos, this chapter will explore how invisible infections have shaped human history. In chapter 8, I take a deep dive into paleoparasitology, an exciting field that leverages ancient DNA (aDNA) technology to study parasites, and uncover how they may have played a role in mass

Thing, and many others have all been inspired by the fundamental nature of parasites.

But why?

One reason parasites captivate us is their obligate intimacy. They must live in or on another organism to survive, or else they perish. Another reason is that they possess extraordinary abilities, from literal mind control to near-immortality through suspended animation in their resting stages. In doing so, parasites challenge our very understanding of life itself. When I give public talks about parasites, I often encounter the same misconception: that parasites, because of their bizarre, alien-like behaviors, are evolutionary outliers—freakish exceptions in the diversity of life. But nothing could be further from the truth. In reality, parasitism is by far the most popular ecological niche on Earth, having evolved independently more than two hundred times![1] To put that into perspective, imagine you could hit a reset button on evolution, wiping the slate clean and allowing life to evolve all over again. The odds are overwhelming that parasitism would re-evolve, over and over again. If evolution were a movie played on repeat, parasitism would be the character that always makes a dramatic entrance, never fails to steal the show, and somehow ends up in every sequel. Take any major branch on the tree of life—bacteria, fungi, invertebrates, vertebrates—and you will find a diverging twig that spawns parasitic descendants.

Once biologists realized just how ubiquitous parasitism is, naturally the next question was why? Why does life keep reinventing this strategy? The answer, at first glance, seems obvious: stealing food, energy, and shelter is far easier than finding it yourself. But parasitism comes with its own unique set of challenges—so much so that one might wonder whether "free" stuff is actually free. For a parasite to succeed, it must overcome two major barriers, or what I refer to as landscapes. The first is the *ecological landscape*: The parasite must navigate the outside world and successfully move from one host to the next—right place, right time, or it dies. The second is the *host landscape*: Once inside a host, the parasite must survive the sheer brutality of its new environment. This includes the host's immune system—a relentless army of specialized cells and cellular soldiers designed to eliminate invaders. Many parasites, especially those that are transmitted *trophically*—i.e., through ingestion by a host—may find themselves in an unpredictable war zone of digestive enzymes and mechanical shearing forces: think of the peristaltic movement of the gut after you consume a mouthwatering ribeye steak. Relatedly, they must be able to either withstand or somehow evade harsh chemical environments found in different parts of a host, such as the highly acidic environment of some species' stomachs. In humans, for instance, the pH of stomach acid in a fasting state is 1.5—only slightly less acidic than battery acid, which is a stinging 0.8. Yet, despite these challenges, parasites thrive. If parasitism is so risky, one might expect organisms to abandon it over time. But the evidence tells a different story. While parasitism has evolved independently more than two hundred times, reversals (where parasites evolve back into free-living organisms) are extremely rare.[2] Yes, life prefers to be parasitic.

In science, disagreements over concepts, ideas, and even basic terminology are the norm rather than the exception. Take the concept of a species, for example. One might assume it has a straightforward definition—after all, it is the fundamental unit of biodiversity, with critical implications for conservation, ecology, and systematics (the discipline concerned with classifying life). We frequently hear that a new species has been discovered or that x number of species are at risk of extinction due to climate change. Yet, biologists continue to debate what a species actually is. Despite countless peer-reviewed papers, spanning from Darwin's time to the present molecular era, there is still no universally accepted definition that satisfies everyone.

The same philosophical and scientific ambiguity applies to parasites. The textbook definition of parasitism describes it as a symbiotic relationship in which one organism benefits at the expense of another, often summarized as a +/- (plus-minus) interaction. A key implication of this definition is that a parasite does not necessarily need to feed directly on its host—it simply must engage in an intimate, obligate association that incurs a fitness cost to the host. I bring this up because throughout this book, I want readers to think of parasitism as a spectrum rather than a rigid, narrowly constrained lifestyle. For example, in my lab, we study tiny polychaete worms that burrow into the shells of mollusks, such as oysters and scallops, where they reside permanently. These worms do not invade the mollusk's tissues, yet they compromise the integrity of the shell, forcing the mollusk to divert energy from growth to shell repair. Instead of feeding on the host, they extend a pair of delicate, tendril-like tentacles (palps) through the shell openings to filter-feed on suspended algae. By applying the textbook definition of parasitism, these worms qualify as parasites, despite never physically feeding on their hosts.

Yes, this book is about parasites—but not in the way most people think about them. Beyond their grotesque and fearsome reputation, parasites are among the most powerful evolutionary forces in nature. They shape ecosystems, drive genetic innovation, and even steer the fate of entire species. Take, for example, the microparasite *Yersinia pestis*, the bacterial agent responsible for the Black Death, one of the deadliest pandemics in human history. This intracellular parasite (a pathogen that can live inside host cells) did not just decimate medieval Europe; it fundamentally rerouted the evolutionary trajectory of human populations living there. The pandemic, which wiped out up to 50 percent of Europe's population between 1347 and 1351, created a bottleneck effect, eliminating individuals who were genetically more susceptible while favoring those with immune-related genetic advantages. Recent genetic studies have revealed that the Black Death exerted some of the strongest selective pressures ever recorded on human DNA.[3] In evolutionary terms, selective pressure refers to the environmental challenges that determine which individuals survive and reproduce, passing their traits to the next generation. Think of it as nature's ultimate test—only those with the right genetic defenses make it through. Just as predators shape the speed of gazelles by catching the slowest ones, diseases like the plague act as an invisible predator, eliminating those with weaker immune defenses while favoring those with protective genetic mutations. In the case of *Yersinia pestis*, this meant that individuals with certain immune-boosting gene variants had a higher chance of surviving and passing those genes on to their children. Biologists have identified two key genetic factors that played a role in this selection: the ERAP2 gene, which helps immune cells recognize and destroy infected cells, and toll-like receptor (TLR) genes, which regulate the immune system's ability to detect and respond to bacterial infections. Over generations, these advantageous traits became more common in the population, permanently altering the genetic landscape of Europe.

The plague is one of the most notorious examples of a parasite altering the fate of an entire species, but it is far from the only one. Similar cases, some just as dramatic, others subtler, can be found throughout history and across countless forms of life. Recent advances in genomic technologies have allowed scientists to peer into the deep evolutionary scars parasites have left on their hosts, revealing how these microscopic invaders have shaped immune systems, driven genetic adaptations, and even influenced species diversity. From the genetic legacy of *Yersinia pestis* in human populations to the enduring influence of schistosome blood flukes, the etiological agents of schistosomiasis, parasites have played an undeniable role in sculpting life as we know it.

Despite their profound ecological impact, parasites are often treated as inconvenient footnotes in biology courses—existing, as

extinctions and human history. From the co-extinction of parasites alongside their host species to the evolutionary pressures that parasites may have exerted on ancient megafauna, this chapter will reveal how parasites have left their fingerprints on Earth's history. Finally, in chapter 9, I look ahead to the fate of parasites in a rapidly changing world. Climate change, habitat destruction, and modern medicine are pushing some parasites toward extinction—but is that a good thing? Some scientists argue that the loss of parasites could have grave unintended consequences, particularly for ecosystem health and even human immune function. As we explore the ecological roles parasites play, we will also examine a more unsettling possibility: the potential for parasites to be harnessed as tools of biowarfare. In an age of synthetic biology, artificial intelligence, and geopolitical instability, could these ancient hitchhikers be repurposed for modern conflict? Should parasites, long considered pests, actually be preserved as an essential, and potentially dangerous, component of biodiversity?

Each chapter will uncover the hidden role parasites have played in shaping the world as we know it, from ancient mass extinctions to human civilizations, from genetic theft to mind control. Through this journey, I hope to convince you that while parasites may be horrifying, grotesque, and even unsettling, they are also among nature's most powerful evolutionary forces. Without them, life on Earth would have followed an entirely different path. And perhaps, just perhaps, by the end of this book, you might even come to admire them.

So, let's begin.

1

PARASITES AS EVOLUTION'S ENGINEERS

Nature is not cruel, only pitilessly indifferent. This is one of the hardest lessons for humans to learn. We cannot accept that things might be neither good nor evil, neither cruel nor kind, but simply callous, indifferent to all suffering, lacking purpose.

—Richard Dawkins, *River Out of Eden: A Darwinian View of Life*

I have always been a fan of the horror genre, whether in books, movies, video games, you name it. In high school, I devoured the works of H. P. Lovecraft, Stephen King, and Clive Barker, and I binged every eerie frame of John Carpenter's films. More recently, one series stood out to me for how it portrayed something I now study professionally: parasites. Released in 2014, *The Strain*, created by Guillermo del Toro and Chuck Hogan, is easily one of the most compelling depictions of parasitic infection I have seen on screen. If you have not watched it, I strongly recommend you do.

Without spoiling too much, the show begins with a mysterious plane arriving in New York City, its passengers seemingly dead. A few "wake up" during a CDC inspection and begin exhibiting fevers, loss of appetite, an eerie buzzing in the ears, and, over time, more extreme anatomical changes such as reabsorption of genitalia. The cause? A transparent, wormlike organism that burrows under the skin, leaving visible trails as it moves deeper into its host. The unsettling part is that this kind of migration is real: certain parasitic worms, such as tropical hookworms, can penetrate the skin and cause a condition called *cutaneous larva migrans*, in which juvenile worms leave itchy, winding tracks under the skin, a traveler's worst nightmare.[1] In *The Strain*, these worms do more than irritate; they hijack their hosts' brain, transforming them into bloodthirsty, zombielike beings.

The parasite in the show seems to borrow features from several real-world nematode worms like *Ascaris*, *Necator americanus* (hookworm), and *Dirofilaria immitis* (dog heartworm). The catch? It violates the "parasite playbook" in a major way: The adult stages are immortal, and it kills far more hosts than it successfully infects. This makes for good television but is a terrible strategy for a real parasite. In nature, parasites must strike a delicate balance: Harm the host too much, and you risk destroying the very ecosystem you depend on. In *The Strain*, the infected humans eventually farm the uninfected in a dystopian reversal. Thankfully, nature has not yet offered us such a grotesque scenario.

Looking back on my love for these macabre tales, I realized it was not the fear that drew me in; it was the villains, the antagonists: vampires in *Salem's Lot*, extraterrestrial spores in *Invasion of the Body Snatchers*. They were often portrayed as evil, but to me they were something else entirely: survivors, beings driven by adaptation, not malice. That perspective carried over into my fascination with parasites when I encountered them as an undergraduate. I saw them not merely as pests but as evolutionary sculptors, Michelangelos working in the dark, creatures so adept at survival that they reshape not just individual hosts but entire ecosystems.

Today, the biological sciences are starting to take that view more seriously. Parasites are no longer just background characters in the story of life; they are starting to get top billing. In 2015, William Campbell, along with Satoshi Ōmura, received one half of the Nobel Prize in Physiology or Medicine for discovering the antiparasitic drug, avermectin, while Youyou Tou received the second half for her work on novel therapies against malaria parasites.[2] Ivermectin, a derivative of avermectin, targets the ion channels in worm muscle cells, paralyzing them and allowing the host to pass the worms in their feces. More than a billion doses of ivermectin have been administered globally, saving millions of people from blindness and disability, especially in low-income regions.[3] It was a triumph of collaborative, elegant science.

Dr. Campbell's Nobel lecture is available on YouTube and worth every minute, but it is his lesser-known banquet speech that truly captures the spirit of this book. Here is an excerpt from that speech:

> Parasites are not generally regarded as being loveable. When we refer to people as parasites, we are not being complimentary, we are not praising them. We tend to think that a parasite is the sort of person who goes through a revolving door on somebody else's push. This is unfair. It's unfair to real parasites, to the innumerable and influential parasites from the world of nature. Your majesties, your royal highnesses, your excellencies, ladies and gentlemen, it is time for parasites to get a little more respect!"[4]

For me, that speech was not just witty and comedic; it was overdue. Parasites are often used as metaphors for greed or degeneracy and are still somehow viewed as "unnatural," but they deserve better. Yet despite this growing recognition, parasitology remains deeply underrepresented in medical education, especially in the United States. In most American medical schools, parasitic diseases are covered only briefly, if at all, usually bundled into a few PowerPoint slides at the tail end of a general microbiology course. In fact, parasitology is not a dedicated component of the MD curriculum at any of the top ten medical schools in the United States. This can be verified by reviewing the publicly available curriculum outlines posted on each school's official website. As Dr. Daniel Griffin, cofounder of Parasites Without Borders, has said, "The subject of parasitology has been essentially marginalized at most medical schools."[5] That neglect has real-world consequences. Many physicians receive no formal training in diagnosing or treating parasitic infections, even though infections like *Giardia*, *Strongyloides*, and *Toxoplasma* still occur in the United States, particularly among immunocompromised patients and recent immigrants.[6] As a result, some parasitic diseases go undiagnosed or misdiagnosed, even in major hospitals.

There is a quiet irony to all this neglect. The very symbol of American medicine, the Rod of Asclepius, a staff entwined with a single serpent and adopted by the American Medical Association, may depict a parasite.[7] One plausible origin of the image traces back to the ancient medical treatment of *Dracunculus medinensis*, the Guinea worm—a parasite that will be covered in more detail later in the book. This parasitic nematode, once widespread in Africa and Asia, emerges painfully from the skin and must be slowly pulled out by winding it around a stick, a technique that has been practiced for thousands of years. Some historians believe this harrowing yet familiar visual, a worm curling around a rod, may have inspired the enduring emblem of healing. If true, then parasitology does not just deserve a place in modern medical education; it helped shape its very symbol.

The contrast with veterinary education, on the other hand, is striking. In most veterinary schools, parasitology is required and rigorously taught as part of the curriculum, complete with lab training, diagnostic techniques, and detailed life cycle studies.[8] This is because parasitic infections are an everyday reality for livestock, pets, and wildlife, and veterinarians are expected to recognize, prevent, and treat them. While future human doctors may graduate having never been tested on a single helminth or protozoan, future veterinarians leave school with a working knowledge of dozens. It is an ironic but telling disparity: When it comes to parasites, your dog's vet is better equipped than your own physician.

This blind spot contrasts sharply with the renewed visibility of parasites in public discourse and pop culture, on our screens, in Nobel ceremonies, and increasingly in scientific research. These hidden invaders have evolved some of the most exquisite biological adaptations in the natural world. They both live in the shadows and shape them. In the rest of this chapter, I explore some of those adaptations. From the jungles of South America to the frozen poles, from microscopic hitchhikers to inch-long worms, parasites can be found wherever life exists. They are not just surviving; they are transforming their hosts, and in doing so, the biological world itself.

PARASITES AS EVOLUTIONARY FORCES

As mentioned in the introduction, parasites are organisms that live in an intimate, and often obligate, association with another organism: the host. In this relationship, one party benefits (+) while the other is harmed (–). This makes parasitism a form of symbiosis, though not a mutualistic one. But the term *harm* deserves clarification. In biology, harm does not necessarily mean sickness or death. Instead, it refers to any reduction in the host's fitness—its ability to survive and reproduce. This harm can take many forms: stolen nutrients, hijacked cellular machinery, suppressed immunity, or even behavioral manipulation. What unites all parasites is their ability to extract resources in a way that benefits their own survival and reproduction, but at a cost to the host's energy budget, longevity, or reproductive output. Yet being a successful parasite involves more than just exploitation. It requires a delicate evolutionary balancing act between two critical aspects of infection: pathogenicity and virulence.

Pathogenicity refers to a parasite's capacity to cause disease in a host: either it is pathogenic or it is not. For example, *Trypanosoma brucei*, the parasite responsible for sleeping sickness, is clearly pathogenic. In contrast, *Lactobacillus acidophilus*, a beneficial gut microbe, is not. Virulence, on the other hand, refers to the severity of the disease caused. While the two terms are related, they are not synonymous. A helpful case study here is *Entamoeba histolytica*, a parasitic protozoan that is definitively pathogenic, yet up to 90 percent of individuals infected with it remain asymptomatic carriers.[9] In a minority of cases, however, the parasite invades the intestinal lining, causing dysentery or liver abscesses, extremely virulent and sometimes fatal outcomes. Thus, while *E. histolytica* is always pathogenic, its virulence can vary dramatically between strains and hosts. One of my students once offered a fitting analogy: "If pathogenicity is whether someone is a criminal, virulence is the length of their rap sheet."

Imagine you are a parasite. You have invaded a host, and you are tapping into their nutrients and energy. You are almost guaranteed to cause harm. But hosts do not take this lying down. If your life cycle involves entering the bloodstream or organs, you risk detection by the immune system. Once the host's defenses are activated, symptoms like fever, inflammation, or fatigue may spike, not necessarily due to your actions, but as collateral damage from the host's immune response. In this sense, virulence can be a byproduct of the host's attempt to fight you off.

The "perfect" parasite is therefore one that strikes a balance: It is pathogenic but minimally virulent. Its life cycle is finely tuned to invade, migrate, feed, and reproduce, all while flying under the radar. It stays just below the immune system's threshold, extracting what it needs without tipping the host into collapse.

During Halloween season, I like to unsettle my students with this unnerving thought: *Any one of you could be infected with a tapeworm right now—and you'd likely have no idea unless it was discovered accidentally during a colonoscopy.* That line always lands well with Gen Z, especially in an age of fourteen-second TikTok videos designed to stoke anxiety. But behind the theatrics, there is a kernel of truth. Tapeworms are, in many ways, the quintessential parasite. Take the pork tapeworm, *Taenia solium*, for example. Its life begins in domesticated pigs, where it exists in a dormant cyst form embedded deep in muscle tissue. When humans eat undercooked pork, perhaps a slightly-too-pink strip of bacon, they risk ingesting one of these cysts. Inside the human gut, the cyst undergoes *excystation*: Its head, armed with hooks and suckers, emerges and latches onto the intestinal wall. There, it resists the constant churning and shearing forces of the digestive system with ease. Once attached, it does not chew or digest. Instead, it passively absorbs nutrients, such as glucose, amino acids, vitamins, and minerals, through its skin-like outer surface.[10] As long as the host keeps eating nutrient-rich meals, the tapeworm might as well have won the lottery. It begins to grow, adding segment after segment, until it can reach an astonishing twenty to twenty-three feet in length inside of you![11] Each segment contains both male and female reproductive organs, allowing it to reproduce without a partner. Its eggs, shed constantly through the host's feces, can go on to infect the next host, completing the cycle again and again. Even more remarkable, the pork tapeworm can live for decades inside a person, existing silently, efficiently, and almost entirely undetected. It does not trigger an aggressive immune response because it resides beyond the reach of most immune surveillance. It is, quite literally, the perfect houseguest, from its own perspective of course. If there were a universal law of parasitism, it might be: *Thou shalt not kill thy host.* But like all so-called universal laws, there are notable exceptions, and we will meet some of those soon enough.

While tapeworms might seem enormous in the world of parasites, they are just one example of the astonishing physical diversity that parasites display. Like all life-forms, parasites come in a vast range of sizes and shapes, and because of their evolutionary success, they have managed to infiltrate nearly every tissue, cavity, and crevice in the bodies of their unsuspecting hosts. To better appreciate their variety and to follow the terminology used throughout this book, it helps to distinguish parasites into two broad categories: microparasites and macroparasites.

Microparasites are the smallest players in this hidden world. These include viruses, bacteria, fungi, and protozoans, organisms so tiny that most can only be seen with high-powered microscopes. What defines them is not just their size but also how they behave inside their hosts. Microparasites typically replicate within the host, often completing several generations over the course of a single infection. Some even live inside the host's cells, hijacking the machinery from within. Macroparasites, in contrast, are large enough to be seen with the naked eye. These include worms like tapeworms and roundworms, as well as external parasites like ticks, lice, and mites. Unlike microparasites, macroparasites do not multiply inside their final host. The number of parasites a host carries, what we call the parasite burden, depends on how many are acquired from the environment or other hosts, not on internal replication. Infections by macroparasites are often chronic, with damage accumulating over time. Among macroparasites, we can make a further distinction: Ectoparasites live on the outside of their hosts (think fleas, ticks, and leeches), while endoparasites live inside, such as tapeworms in the gut or filarial roundworms in the bloodstream. This basic framework—micro versus macro, endo versus ecto—will appear often throughout the chapters ahead. But as we will soon discover, parasites are notorious for breaking the rules, and many blur the lines between categories in fascinating and sometimes horrifying ways.

Parasitism is not some fringe lifestyle; it is one of the most ancient and widespread strategies in the history of life. In fact, some biologists estimate that as many as half of all species on Earth live as parasites at some point in their lives.[12] Fossil evidence suggests that parasitism as we understand it emerged more than five hundred million years ago, likely evolving independently dozens of times across unrelated lineages. Flatworms, mites, fungi, wasps, and even flowering plants have all stumbled upon the same winning idea: Why struggle to survive in a chaotic environment when you can outsource your survival to another organism? Parasitism works because it taps directly into the energy flow of ecosystems. Every living creature is a bundle of nutrients: prepackaged biological capital. Rather than expend resources hunting prey or producing their own food, parasites simply move in and exploit the systems that are already in place. This strategy is energetically efficient, evolutionarily flexible, and, when executed properly, remarkably sustainable. Its ubiquity is also a testament to its success. There are parasitic bacteria infecting deep-sea vent organisms, parasitic protozoa that have evolved alongside mammals for millennia, and parasitic wasps that can override the immune systems of their insect hosts with biochemical precision. Parasites wind through the branches of the tree of life like ivy, shaping the very structure of ecosystems.

Because parasites extract resources from their hosts and often cause disease, they exert intensive selective pressure, making them agents of evolutionary change—they are not just along for the ride. Hosts that survive or resist infection gain a reproductive advantage, leading to the evolution of immune systems, defensive behaviors, and even physiological traits that reduce vulnerability to parasitism. In fact, some of evolution's most iconic features may have emerged because of parasitic threats. The complexity of the vertebrate immune system? Likely a response to a parasitic challenge. The behavior of grooming in primates? A defense against ectoparasites like lice and ticks. Even sexual reproduction, according to the Red Queen hypothesis (discussed briefly in the next section and in more detail in chapter 5), may have evolved as a way for hosts to stay one step ahead of rapidly evolving parasites, shuffling their genetic deck to produce offspring with new combinations of resistance traits. On the flip side, parasites also evolve in response to their hosts. The result is a coevolutionary arms race in which adaptations in one species drive counteradaptations in the other. This evolutionary back-and-forth can shape not only individual species but entire communities and ecosystems. In some cases, parasites even manipulate host behavior to increase their transmission rates—a phenomenon so bizarre and precise that it challenges our very definition of agency in nature.

ARMS RACE AND ADAPTIVE SCULPTING

This is the perfect place to introduce the Red Queen hypothesis, which elegantly captures the evolutionary tug-of-war between parasites and their hosts. First proposed by the evolutionary biologist Leigh Van Valen in his 1973 paper "A New Evolutionary Law," the hypothesis takes its name from a scene in *Through the Looking-Glass* by Lewis Carroll in which the Red Queen tells Alice, "It takes all the running you can do, to stay in the same place."[13]

In evolutionary biology, the Red Queen dynamic suggests that species must constantly adapt and evolve, not to get ahead but simply to keep up with their coevolving rivals. Nowhere is this more evident than in the relationship between parasites and hosts: As parasites evolve new ways to exploit their hosts, hosts must evolve countermeasures just to survive. The result is an ongoing arms race, a cycle of adaptation and counteradaptation in which neither side gains a permanent advantage.

Van Valen analyzed fossil data from a wide range of taxa, including mollusks, mammals, and others, and observed that the probability of extinction of a group was constant over time, regardless of how long that group had existed. In other words, older lineages were no more or less likely to go extinct than younger ones. This was surprising. If species were improving over time through adaptation, you would expect older species to become more resilient and less likely to go extinct. Instead, Van Valen found that extinction rates did not decrease with evolutionary age, implying a kind of "moving target" environment—an arms race, not a finish line. To explain this pattern, Van Valen hypothesized that evolutionary success is relative, not absolute. Even if a species adapts and improves, it still must deal with other evolving species such as predators, prey, competitors, and especially parasites and pathogens, which can evolve very rapidly. In the context of parasitism, this is a perpetual game of catch-up: Hosts evolve new defenses and parasites evolve to bypass them.

This relentless pressure often produces surprising biological innovations. For example, consider immune system complexity. The vertebrate immune system, with its dazzling array of specialized cells and molecules, may have evolved largely in response to the threat posed by rapidly evolving parasites.[14] Even basic physical traits, like thicker shells in mollusks or tougher exoskeletons in insects, can be traced to parasite-driven selection. One striking example comes from the evolutionary duel between freshwater snails and their trematode (flatworm) parasites. In many species, the shape of a snail's shell—its thickness, color, and even the direction of its spiral—can influence how easily a parasite can invade.[15] Trematodes typically enter through the soft tissue near the shell's opening, but not all shells offer the same welcome. Some are thicker, some have narrower or more tightly coiled openings, and others spiral the opposite way entirely. These subtle differences can make it physically harder for parasites to get inside, and in some cases, the shell itself can trap or wall off invading larvae. Because parasites tend to exploit the most accessible hosts, snails that differ slightly in shell design may enjoy a temporary advantage, simply because their parasites are not well adapted to them. Over time, this tug-of-war can help maintain shell diversity within a population—a process biologists call frequency-dependent selection, in which rare traits are favored precisely because they are rare. In this way, the parasite becomes an invisible sculptor, shaping not just the fate of individual snails but the very contours of their shells.

Parasites also influence life history strategies of species. Life history strategies refer to the timing and pace of an organism's development and reproduction. In host populations facing intense parasitic pressure, hosts may reproduce earlier, even if it means producing fewer and lower-quality offspring. The logic is simple: Reproduce before the parasite can cripple or kill you. For example, consider parasitoid wasps. Unlike conventional textbook parasites, parasitoids use a host, usually the larval form of another insect, as a living food source for their developing young, and the relationship always ends in the host's death. In many cases, a female wasp deposits her eggs on or inside an unsuspecting caterpillar or insect larva. Once the eggs hatch, the larvae grow by feeding on the host from within, carefully consuming nonvital organs first to keep it alive as long as possible. It is a slow, brutal, and unforgiving coup de grâce—something you would not wish on your worst enemy. In some insect communities, the presence of these parasitoid wasps causes hosts to accelerate their metamorphosis, leaving their vulnerable larval stage earlier than normal. This comes at a cost of course—smaller adult size, shorter lifespan—but it can mean the difference between life and death.

Behaviorally, some species go to great lengths to avoid parasites altogether. Mammals groom obsessively to remove ectoparasites. Frogs learn to avoid egg-laying sites infested with parasitic fly larvae. Even plants have been observed to alter their leaf chemistry or flowering times in response to parasitic threats. In all these cases, parasites act not merely as passengers or background noise in the evolutionary story, but as sculptors, chiseling away at the genetic marble of their hosts, reshaping forms, behaviors, and strategies in their own image. And while many of these adaptations are defensive, they also generate new diversity, pushing evolution in unexpected directions. As we will see in later chapters, some of the most bizarre and beautiful biological phenomena emerge because of parasites.

BODY SNATCHERS: MORPHOLOGICAL INNOVATION AND CONTROL

My most treasured childhood memories revolve around the ocean. Growing up in the island nation of Trinidad, nearly every child had some connection to the sea—cookouts on the sand, snorkeling adventures, or fishing trips with family. For me, it was the surreal sensation of submerging just beneath the surface and watching the sunlight filter down through the water. For those brief, breathless moments, I felt transported to another world—alien, vibrant, and filled with strange and wondrous life.

One memory from that time quietly changed the course of my life. It was a Sunday evening in August, just before the start of my sophomore year in high school. My father and I were headed to Toco, a rural fishing village on the northeastern coast of the island. Even now, Toco remains a sleepy place, dotted with coconut palms, thick forest, and a patchwork of homes with galvanized roofs. Kids played cricket in the streets while adults collected water from communal standpipes. The village always felt timeless—rooted in the land, the sea, and the rhythms of nature.

We crested a hill just past a sign that read "Salybia Beach," and there it was—the sharp, intoxicating scent of the ocean. It is a smell burned into my brainstem, one that still triggers nostalgia today, even though the beach has become more of a second office than a place of recreation. It was low tide when we arrived. Most families were packing up for the day, save for a few who stayed behind to cook dinner over driftwood fires, a beach tradition known in Trinidad as a *river lime*. My father parked his treasured Suzuki Jimny 4x4 carefully away from the palms, and I bolted toward my secret hideaway.

"Andy! Be careful! Don't go too far out!" he called after me, as I sprinted across the sand.

Truth be told, I have never been much of a swimmer—something I hesitate to admit as a marine biologist—and my father knew this and always kept a wary eye on me, especially with the tide soon to return.

My secret was a little-known reef tucked away from the main swimming area and out of the lifeguards' sight. I stripped down to my trunks, dropped my clothes near an old shack, and dashed into the shallows. The reef was alive with color and movement: Anemones waved their tentacles, hermit crabs scuttled between rocks, and small fish darted through the water. I carefully avoided stepping on the finger corals, which carpeted the seabed like a fragile mosaic.

Then I spotted a group of crabs, no bigger than the palm of my hand, grazing on algae atop a boulder. One crab stood out. It looked like the others, same size, same color, but something was off. Its rear end was bulging, oddly swollen. Curious, I gently picked it up and turned it over, expecting to see eggs tucked beneath its abdomen. But instead of neat clusters of orange spheres, there was a yellowish blob—a soft, lumpy pouch with tiny pores or filaments. The crab did not thrash or try to pinch me. It just sat there in my hand, oddly calm, as if in a trance.

I had no idea what it was, but I was both disturbed and captivated. I begged my father to let me take it home, imagining grand experiments I would conduct in my bedroom. He was not amused. I returned the crab to its rock.

It was seven years before I encountered that yellow blob again—this time as an undergraduate, on a field excursion to the rocky shores of Long Island. Only now did I learn what I had really seen: the reproductive organ of a parasitic barnacle. Specifically, it was the "externa" of a rhizocephalan barnacle—a grotesque structure that protrudes from the host crab's abdomen. Inside the crab, hidden from view, was the interna: a sprawling network of rootlike filaments that had infiltrated the crab's tissues, organs, and even its nervous system. The crab was no longer truly itself. It had become a living incubator, its behavior and body repurposed in service of the parasite.

In male crabs, rhizocephalans trigger astonishing morphological changes. The narrow, semicircular abdomen becomes broad and rounded—reshaped to mimic a female's, so it can better cradle the externa. Reproductive organs shrink or disappear entirely. The gonads degenerate. The male begins to perform female-like behaviors: abdomen-waving, nurturing movements that normally accompany larval release. Yes—male crabs can be feminized.[16] And not just externally. These parasites hijack hormonal pathways, reshaping not only the host's anatomy but also its behavior. The crab becomes a surrogate parent, tending to the parasite's brood as if it were its own. Female crabs are not spared. When they are infected, the rhizocephalan occupies the space that would normally carry an egg mass. The host stops reproducing. Her ovaries atrophy. But her maternal instincts remain—redirected toward the parasite's offspring, which she protects and oxygenates just as she would her own embryos. This kind of manipulation is known as parasitic castration, and in rhizocephalans, it is absolute. The crab no longer eats, mates, or behaves like a normal crab. It becomes a hollowed-out husk with one evolutionary function left: to raise the parasite's young.

And here's the most chilling part: The parasite is a barnacle.

Despite its wormlike appearance inside the crab, Rhizocephala belong to the same group of crustaceans that includes gooseneck and acorn barnacles, those sharp crowns of calcium carbonate that we see dotting rocky shores on the beach. They evolved from free-living filter feeders, but over time they abandoned their shells, their limbs, and even their resemblance to animals. The adult rhizocephalan no longer looks like a barnacle. It is a network of tissue and a reproductive pod—a parasite so reduced and specialized that it is more virus than animal.

In the evolutionary arms race between parasites and hosts, rhizocephalans are body snatchers *par excellence.* They do not just live inside their hosts; they become them, turning crab into cradle, instinct into puppet strings.

Zombified crabs are easy enough to spot. After all, it is hard to miss a giant yellow sac bulging from the rear of a crab. Rhizocephalan externae are conspicuous—once you have seen one, you start spotting them everywhere. But not all parasite-induced anatomical changes are so obvious. In many cases, the parasite is microscopic, and the damage it causes can be mistaken for pollution, developmental errors, or genetic defects. That is exactly what happened in the summer of 1995, when a group of seventh graders from Minnesota New Country School, led by teacher Cindy Reinitz, went on a field trip near the small town of Henderson. The class was monitoring frog populations as part of an environmental biology project. While collecting live frogs from a local pond, the students made a macabre discovery: The frogs were malformed. Some were missing limbs entirely; others had extra legs, too many toes, or twisted, useless appendages.[17]

In a 2021 interview with FOX 9 News, one of the students, Reta Lind, recalled her initial reaction:

> "Our brains went, 'Oh, what is this?'"
>
> "It wasn't anything gross," she added. "It was more like, 'This is strange. I want to see what's going on.'"[18]

That spark of curiosity is the essence of science. But while the kids were fascinated, many adults were alarmed. Photos of the deformed frogs circulated quickly, and what started as a classroom project exploded into a national controversy. Environmentalists blamed pollution. Politicians demanded investigations. News outlets used the word *mutant* with unsettling frequency, raising fears about radiation, pesticides, and industrial runoff. Was this a sign of a poisoned world?

The Minnesota Pollution Control Agency, along with the Environmental Protection Agency (EPA), launched investigations. There were calls for stricter wetland protections and tighter regulations on agriculture and industry. Meanwhile, conservative lawmakers and agribusiness lobbies pushed back, fearing that frog deformities would be used as a political weapon to restrict chemical use.

I will not go into the full details of the political drama here—though for those interested, I strongly recommend Judy Helgen's *Peril in the Ponds*.[19] As a state biologist at the time, Helgen offers a gripping insider account of the scientific and bureaucratic turmoil that followed.

It was not until 1999 that the mystery began to unravel. That year, Pieter Johnson, then at Stanford University, and his colleagues published a landmark paper in *Science* that solved the puzzle. In their study, they exposed tadpoles to varying doses of the larval stage of a parasitic flatworm—a trematode called *Ribeiroia ondatrae*. The results were stunning: The tadpoles developed extra limbs, skin webbings, and malformed legs, almost identical to the deformities seen in the field. Their earlier field surveys had already found encysted parasite larvae (metacercariae) embedded in the limbs of wild frogs, as well as in the snails that shared their ponds. The lab experiments confirmed causation: *Ribeiroia* was the culprit. In their study, the researchers found that timing was everything: Tadpoles were most vulnerable to deformity only during a narrow window of limb development.[20]

But this raises an evolutionary question: Why would a parasite deliberately cripple its host? Here is where things get clever. *Ribeiroia*, like many trematodes, has a complex life cycle that involves multiple hosts. Its final destination—the only host in which it can reproduce sexually—is not the frog but a waterbird, such as a heron or an egret. Adult parasites in the bird's digestive tract produce eggs that exit the host in feces and end up in nearby ponds. Once in the water, the eggs hatch into miracidia—free-swimming larvae that live only a few hours and cannot feed. They must locate and infect a snail, their first intermediate host, or die trying. To do this, miracidia are armed with chemosensory cilia, allowing them to follow the trail of snail mucus and metabolic byproducts—an evolutionary form of scent-tracking known as chemotaxis. They are also positively phototactic, swimming toward light, which brings them to the shallow, sunlit waters where snails graze on algae. If successful, the miracidium burrows into the snail and begins an asexual reproduction phase. Within the snail, the parasite transforms into cercariae—larger, more muscular larvae with powerful tails. These are released back into the water, where they must find a tadpole, their second intermediate host. Like their predecessors, cercariae use a mix of chemical and tactile cues to find and infect their target. But they do not invade just anywhere—they preferentially encyst near the developing limb buds. This is where the real disruption happens. Encystment around the limb buds interferes with the local signaling pathways—including FGF and sonic hedgehog (yes, the actual name of the signaling molecule)—that guide proper limb formation. As a result, the growing tissue becomes disorganized, leading to frogs with extra legs, fused limbs, or digit abnormalities. These deformities are not genetic; they are the product of mechanical and developmental interference during a critical window of embryogenesis.

And now comes the obvious kicker: Deformed frogs are easier for birds to catch. Their swimming is impaired, their reflexes slowed.

This increases the likelihood that they will be eaten, completing the parasite's life cycle by delivering it into the digestive tract of its definitive host.

In short, *Ribeiroia* does not hijack its host's behavior the way rhizocephalans do. Instead, it disables the host, sacrificing it to predation in a grisly but effective strategy of transmission—not subtle, not elegant, but it works. And in the cold calculus of evolution, that is all that matters.

If the idea of a parasite turning a crab into a brooding machine or sprouting legs on a frog was not disturbing enough, this last case pushes the concept of morphological control into unsettling new territory. What if the parasite does not just hijack the host, but transforms into something else entirely inside it? In 2013, a forty-one-year-old man in Medellín, Colombia, walked into a clinic with symptoms that were not unusual for someone with advanced HIV: fatigue, weight loss, coughing, and enlarged lymph nodes. He had been living with HIV since 2006 and was severely immunocompromised. Scans revealed lung nodules, liver and adrenal abnormalities, and widespread lymphadenopathy. Doctors feared the worst—disseminated cancer. Biopsies of the lymph nodes showed masses of small, undifferentiated cells invading the tissue. Under the microscope, they had all the features of cancer: monomorphic nuclei, mitotic figures, syncytia, and cytologic atypia. But something was off. The cells were too small—even smaller than human red blood cells—and they did not behave like any known human cancer. Molecular tests soon delivered an astonishing result: The DNA inside the "tumor" cells was not human! It belonged to a tapeworm.[21] Specifically, the culprit was *Hymenolepis nana*, the dwarf tapeworm—the most common human tapeworm in the world, infecting an estimated seventy-five million people globally. In most cases, it causes no symptoms. Unlike its distant cousin, the pork tapeworm, *H. nana* is unique in that it can complete its life cycle entirely within a single human host, without needing an intermediate host. In children or immunocompromised individuals, it can autoinfect, leading to persistent, high-burden infections.

But this case was something entirely new. The tapeworm cells invading the man's lungs and lymph nodes were not organized as a worm—they lacked any recognizable parasite structures. Instead, they had mutated into a neoplastic mass. Genetic sequencing showed structural genomic rearrangements, copy-number amplifications, and insertional mutations—the same types of genetic chaos seen in human cancers. And just like cancer, these tapeworm-derived cells were invading tissue, spreading, and replicating uncontrollably.

How could this happen?

The answer may lie in the parasite's stem cells. Tapeworms, like other flatworms, rely on pluripotent stem cells called neoblasts to regenerate tissue. In healthy hosts, the parasite's development is tightly regulated. But in an immunosuppressed environment, such as that of this patient, the normal checks and balances break down. Without immune surveillance, mutant neoblasts may accumulate unchecked mutations, leading to a rogue, self-replicating population—a parasite-derived cancer growing inside the host.

To be clear, this was not a case of the host developing cancer because of the parasite—like liver flukes causing bile duct cancer, or HPV triggering cervical tumors. In this case, the parasite itself *became* the cancer. It is, to date, the only known example of a multicellular parasite transforming into a malignant entity inside a human being. And perhaps most unsettling, because the cells originated from *H. nana*, they were not recognized as foreign by standard diagnostic tools. They did not look like a parasite, and they did not look like a human tumor. They were something in between—a chimeric threat that blurred the line between infection and neoplasia.

The patient died before a definitive diagnosis was made. Only in the final days of his life did scientists at the CDC confirm what they were seeing. No treatment was available. Even standard antiparasitic drugs like albendazole—effective against intact worms—show little promise against clonal proliferations of tapeworm stem cells.

This case leaves us with a haunting possibility: What other parasites, under the right (or wrong) conditions, might evolve into something else entirely—something that defies both biology and taxonomy? Parasites, it seems, are not just body snatchers. Sometimes, they become the body itself.

LIFE CYCLE COMPLEXITY: THE PARASITE'S GRAND DESIGNS

The early months of the COVID-19 pandemic now feel like a surreal blur—a collective trauma frozen in time. But for those who lived through it on the front lines, it was no dream. In March 2020, as lockdowns spread across the United States and much of the world, a new and poorly understood pathogen, SARS-CoV-2, was racing through human populations. In New York City, hospitals were overrun. Refrigerated trucks lined up outside medical centers to serve as temporary morgues. My best friend, Charmaine, a nurse at New York–Presbyterian Hospital, was overseeing a COVID unit during the worst of it. My mother, a certified nursing assistant in Queens, sent me regular updates about the dire conditions in her own senior facility. Thousands of miles away, in India, the crisis took on a similarly haunting scale—funeral pyres burned in parking lots as crematoriums overflowed. Amid the global chaos, one of the first tasks for scientists and epidemiologists was to figure out the life cycle of the virus. How did it transmit? Was it strictly human-to-human, or were there animal reservoirs? Could it linger on surfaces? Did it spread through aerosols? Without answering these basic questions, public health interventions such as quarantines, travel bans, and surface disinfection were being made in the dark. Only by understanding where the virus lived, how it moved, and what it required to reproduce could scientists offer guidance to break its transmission chain.

This challenge is not unique to viruses. In parasitology, knowing the life cycle of a parasite is fundamental. In fact, my students quickly learn that if they want to pass my exams, they better know their life cycles inside out. We spend hours diagramming the journey from egg to larva to adult, through host after host, because each stage can be thought of as a potential point of intervention. I often tell them if a parasite's life cycle is a story, it is one with a very specific plot. There is a precise order in which the events must unfold for the next generation to survive. Think of it like an electrical circuit: Each node represents a transformation or transmission point, and if even one link fails, the whole system collapses. From a public health standpoint, knowing which "switches" to flip, or break, can mean the difference between containment and outbreak. Some parasites have simple, monoxenous life cycles, meaning they require only a single host. Take *Ascaris lumbricoides*, the common human roundworm: It infects the intestine, matures, and lays eggs, and the cycle continues when those eggs are ingested again. Clean water and sanitation can break the loop. But others have complex, heteroxenous life cycles, requiring two or more hosts, often of different species. These include the pork tapeworm discussed earlier, which uses pigs as intermediate hosts and humans as definitive hosts. In such cases, control becomes more difficult because multiple environments, and species, must be managed simultaneously.

So far, we have discussed intermediate and definitive hosts as the two main types of hosts that parasites use to complete their life cycle; however, some parasites have additional hosts, which adds to the complexity of their life cycles. Paratenic hosts, also known as transport hosts, are hosts that are not required for the parasite to complete its life cycle but can carry the parasite in an infective form, often bridging the gap between intermediate and definitive hosts. Think of a paratenic host as a rideshare service for the parasite, a way to hitch a ride without paying a fare. The parasite gets in, stays quiet, and waits for the host to be eaten by the "right" host to continue its journey. Nothing changes in the parasite while it is in the paratenic host; it is just biding its time in a biological waiting room. For example, *Toxocara canis*, the dog roundworm, can end up in a mouse, which is not part of its normal development. The larvae do not grow or reproduce there—they just hang out. But if a dog eats that mouse, the parasite reaches the gut and resumes development into an adult worm. In this scenario, the mouse is a paratenic host—a detour, not a destination.

Accidental or dead-end hosts are not part of a parasite's canonical life cycle, but they can still become inadvertently entangled in transmission routes. In these hosts, the parasite is unable to complete development or reproduce, effectively reaching a biological cul-de-sac. It may survive for a time, or even cause disease, but the transmission chain is broken—development halts, and no new generations are produced. Take, for example, the single-celled protozoan *Toxoplasma gondii*. This parasite uses cats as its definitive host—where sexual reproduction occurs—and rodents, such as mice, as intermediate hosts. Infected cats typically show no symptoms but shed microscopic oocysts in their feces. These oocysts contaminate the environment and are then accidentally consumed by rodents. Once inside a mouse, *Toxoplasma* forms tissue cysts, particularly in neural and muscle tissue. If the cat later eats the infected rodent, the life cycle is complete.

But *Toxoplasma* is not picky. Pigs, sheep, and goats can also serve as intermediate hosts. In many parts of the world, poor sanitation or the consumption of undercooked meat allows *Toxoplasma* to spread outside of its intended hosts. In humans, infection can occur by eating tissue cysts in undercooked pork or lamb, or by ingesting oocysts from contaminated water, soil, or food—sometimes due to improper handling of cat litter. Most healthy individuals experience no symptoms. When symptoms do appear, they are typically mild and flu-like: fatigue, swollen lymph nodes, or low-grade fever. But the real concern is during pregnancy. *Toxoplasma* can cross the placenta, a process known as transplacental transmission, and infect the fetus. Depending on when the infection occurs, consequences can range from miscarriage or stillbirth to neurological damage, blindness, or seizures that may not appear until later in life.[22] Worldwide, it is estimated that around 190,000 new cases of congenital toxoplasmosis occur each year, with the highest rates found in Latin America, parts of Africa, and some European countries.[23] In the United States, the estimated prevalence is approximately 0.5 cases per 10,000 live births.[24] Humans are considered dead-end hosts for *Toxoplasma gondii*. Once the parasite encysts in our tissues, the life cycle should, theoretically, stop. But as Jeff Goldblum famously says in *Jurassic Park*: "Life finds a way." When I cover this parasite in my Ecology of Infectious Diseases course, I divide the class into groups and challenge them to consider how *Toxoplasma*'s life cycle might still be completed even if it ends up in a human. It is part dark humor, part pedagogical strategy, but it does not take long before students land on the highly implausible but technically possible scenario: If a human dies while infected and a cat consumes the body soon afterward, the parasite could reenter the cat and resume its normal cycle.

While this is an extreme rarity—and cats are more likely to nibble than feast—*Toxoplasma* cysts do remain viable for a short time after death. So yes, in theory, Mr. Nibbles could serve as the unintentional rescuer of a parasite stranded in a dead-end host.

Is it likely? No.

Is it a little horrifying? Definitely.

Would it make a great horror screenplay? Absolutely. (Cujo meets *The Walking Dead*, with a dash of *Contagion*.)

Another type of host that you may have heard of, especially with regard to diseases like malaria and dengue fever, is a vector. A vector is an organism—usually an arthropod like a mosquito, tick, or flea—that transmits a parasite from one host to another without becoming seriously ill itself. Some vectors simply carry the parasite from one place to another (mechanical vectors), while others are biological vectors, meaning the parasite must develop or reproduce inside the vector's body before it can infect the next host. Think of a vector as a parasite's taxi, but not just any taxi. In many cases, it is a custom-engineered vehicle that the parasite has adapted to, hijacking its biology to get from one host to another with perfect timing. Vectors are not always passive chauffeurs in the parasite's grand design. In some cases, they are more like hypnotized accomplices. Take *Plasmodium*, the protozoan responsible for malaria. It does not just travel inside the female *Anopheles* mosquito; it tweaks her behavior to increase the odds of spreading. Infected mosquitos are persistent, more likely to feed at night when humans are sleeping, and may take longer, more frequent bites, raising the chance of transmission with each blood meal. It is as if the parasite is whispering in the mosquito's ear urging her to seek out blood with greater urgency, all in service of completing its life cycle.

Finally, from a public health perspective, arguably the most important host, if present, is the reservoir. A reservoir is any animal, plant, or biological source that harbors a parasite over the long term and serves as a persistent wellspring of infection for other hosts. Reservoirs often show no symptoms of disease, yet they quietly keep the parasite alive and circulating in the environment, even when primary hosts are absent, immune, or controlled. Think of a reservoir as a savings account for a parasite. Even when the parasite is not actively "spending" (infecting its primary host), it remains safely tucked away, earning "interest" and waiting for the right conditions to strike again. Even if epidemiologists manage to piece together the life cycle of a highly virulent parasite and develop control measures to interrupt its circuitry, all of those efforts could unravel if a hidden reservoir goes undetected. In such cases, the parasite can simply reemerge from its quiet refuge, resetting the outbreak clock and blindsiding public health efforts. For example, during the COVID-19 pandemic, epidemiologists were deeply concerned that SARS-CoV-2 might establish itself in wild or domesticated animal reservoirs. And indeed, infections in mink, white-tailed deer, and even house cats raised alarms about the virus reemerging from nonhuman hosts long after the human population had developed vaccines or natural immunity.[25]

Another example, perhaps more heartbreaking, is the global campaign to eradicate the Guinea worm, *Dracunculus medinensis*. This nematode is one of the most gruesome human parasites in history. It has been known since antiquity, referenced in the Hebrew Bible as a "fiery serpent" and recorded by ancient physicians such as Galen and Pliny the Elder.[26] Its life cycle is elegantly horrifying. Tiny freshwater crustaceans called copepods serve as intermediate hosts, carrying the worm's infective larvae. When a human drinks unfiltered water containing these copepods, stomach acid kills the crustacean and releases the larvae. These then burrow through the intestinal wall, mature, and mate inside the body, making humans the definitive host. The male dies, but the fertilized female embarks on a slow migration through the connective tissues, eventually reaching the lower limbs. There, about a year later, she causes a painful blister, usually on the leg or foot. The agony is so intense that victims are driven to submerge their limbs in water, offering momentary relief, but also triggering the worm to emerge and release thousands of larvae. Those larvae are consumed by copepods, and the cycle begins again. For millions in sub-Saharan Africa, Guinea worm emergence was a season of suffering. Entire communities ground to a halt as victims, with worms painfully wriggling from their skin, were bedridden for weeks. There was no treatment, no vaccine, just slow agony and open wounds prone to bacterial infection. In a world where daily life depended on physical labor, Guinea worm stole the most precious resource of all: time. Children missed school. Farmers missed planting seasons. Families lost income. And communities reshaped their lives around the parasite's grim schedule.

As recently as 1986, an estimated 3.5 million people in twenty-one countries were infected each year.[27] But thanks to massive efforts led by the Carter Center and other organizations, that number plummeted. By 2015, *The Atlantic* ran a hopeful headline: "The Second Human Disease Ever Eradicated . . . Only 126 cases of Guinea worm remain before the parasite disappears from humanity entirely."[28] Just as scientists stood on the brink of eradication, a new twist emerged. Between 2012 and 2016, cases began emerging in domestic dogs in Chad. At first, this seemed anomalous, until researchers realized it was not. In 2020, Robbie McDonald and colleagues published a paper in *PLOS Neglected Tropical Diseases* confirming that dogs were not just accidental hosts, they were reservoirs.[29] This meant the parasite had not disappeared, merely gone underground. The discovery of dogs as reservoirs was not a sign of the parasite adapting to a new host, but rather of scientists finally noticing a transmission pathway that had been hidden all along. The global campaign had been laser-focused on human cases, inadvertently allowing canine infections to slip under the radar. Only when human cases dwindled did this hidden link become visible, forcing scientists to revise what they thought was a fully resolved life cycle.

Reservoirs are now a growing concern in a rapidly changing world. As humans expand into forests, wetlands, and other wild habitats, we increasingly encounter animals that harbor parasites with no previous history in our species. This kind of contact, driven by habitat destruction, urbanization, and global trade, sets the stage for spillover events—when a parasite jumps from a wildlife reservoir into a new, often unprepared host like us. These encounters, which once might have been rare, are now becoming more common.

I return to this in the final chapter, where I explore how human activity is reshaping the global parasite landscape and how parasites, in turn, may be shaping our future.

When students first learn about parasites with multiple hosts, they often assume the hosts were simply available and the parasite made do with what it found. But there is growing evidence that in some cases, the parasite may actually have helped shape the link. In other words, some predator-prey relationships may exist in part because of parasites. Take the trematode *Dicrocoelium dendriticum*, also known by its common name, the lancet liver fluke. Its life cycle is already an elaborate relay: Eggs are shed in the feces of grazing mammals, like cows or sheep. Snails consume the eggs and then release slime balls packed with infectious larvae. Ants, mistaking these slime balls for food, consume them—and become the parasite's next host. But *Dicrocoelium* does not stop at infection. It manipulates the ant's behavior, forcing it to climb up a blade of grass and cling to the tip during the cool hours of morning or evening—prime grazing time for a cow or sheep. Once eaten, the parasite completes its journey to the liver. Without this behavioral nudge, the connection between ant and cow would rarely happen. The parasite has, in essence, created a bridge between two very different creatures. Similarly, *Toxoplasma gondii* does not just infect rodents; it reprograms them. Infected mice lose their fear of cats. Some even seem drawn to cat urine. This bizarre quirk of neurobiology makes it far more likely that a cat will catch and eat the mouse, delivering the parasite back to its definitive host.

PARASITES AS DRIVERS OF SPECIATION

Parasites do not just influence who survives; they also shape who gets to mate. In many animals, traits like bright plumage, elaborate displays, or bold behaviors are energetically costly and can make individuals more visible to predators. Yet these same traits often act as honest signals of health and parasite resistance: Only individuals in prime condition, often those who can mount effective immune responses, can afford to display them. This dynamic forms the basis of parasite-mediated sexual selection, a process by which mating preferences evolve in response to the pressures parasites impose on host fitness. For example, in the three-spined stickleback (*Gasterosteus aculeatus*)—a small, spiny fish found in lakes, rivers, and coastal waters across the Northern Hemisphere—males from different habitats face different parasite communities. These differences in local parasite pressure can drive divergence in traits like coloration and courtship behavior. Females selecting mates with brighter coloration may not just be choosing flashy partners; they may be indirectly selecting for genes that confer resistance to local parasite fauna. Over time, if parasite pressures vary consistently across habitats, these divergent preferences and traits can reinforce reproductive isolation between populations, contributing to ecological speciation.[30]

In this way, parasites quietly shape the evolutionary journey of their hosts. When different environments expose animals to different parasites, those animals adapt not only in how they defend themselves but also in how they choose their mates. A red throat might be irresistible in one pond and meaningless in another. Mating becomes selective, gene pools begin to split, and what began as a difference in immune response may end, generations later, in the birth of a new species. Far from just tearing life down, parasites may help build it up, one evolutionary fork at a time.

ECOSYSTEM ARCHITECTS

Since 2013, thousands of people have taken part in the annual Florida Python Challenge, a state-sponsored competition to remove as many

Burmese pythons from the Everglades as possible. The ten-day event draws both professional wranglers and enthusiastic amateurs, all competing to bag as many of these invasive reptiles as they can. Winners can take home more than $30,000 in prizes, with $10,000 awarded to the person who catches the most. In 2024, Ronald Kiger claimed the grand prize after removing a total of twenty pythons.

Aside from scoring bipartisan political points in Florida's legislature, the Python Challenge brings needed attention to a serious conservation issue in arguably the most biodiverse region of the United States. The Burmese python (*Python bivittatus*) consistently ranks among the most damaging invasive species in North America. It has been linked to 90–100 percent declines in mammal populations in the Everglades, including rabbits, raccoons, opossums, and foxes.[31]

The story of its invasion reads like an ecological thriller. The first documented python in Florida appeared in 1979. The real explosion came in the 1990s, following the high-profile escape of dozens, possibly hundreds, of snakes after Hurricane Andrew destroyed a reptile breeding facility in Dade County. That disaster, combined with frequent pet releases by overwhelmed owners, set the stage for a population boom. Since then, the snake has steadily spread across southern Florida. Species distribution models suggest its range will likely remain confined to South Florida (and possibly extreme southern Texas), due to poor cold tolerance. However, recent genetic evidence hints at natural selection favoring increased cold resistance in some lineages.[32] As a New Yorker and now a New Englander, I can only say this is one problem we can do without, thank you very much.

So, what explains their success? Part of it is ecological fit. The Everglades' warm, wet, vegetated environment is tailor-made for these ambush predators. Burmese pythons are dietary generalists: More than seventy native species have been found in their stomachs.[33] The thick vegetation and vast wetlands provide perfect camouflage, and with no native predators to threaten adults, their survival rate is sky-high. Add to that their rapid reproduction and long lifespan, and you have a recipe for an apex invader.

But there is another, more subtle reason for their dominance—one in which parasites take center stage. This is called the enemy release hypothesis. The idea is that when species invade new environments, they often leave behind the parasites and pathogens that once kept them in check. The result? They face fewer biological constraints and gain a competitive edge over native species. In Florida, studies have shown that Burmese pythons appear to host fewer parasite species than might be expected in their native Southeast Asian range.[34] With fewer internal enemies, these snakes can allocate more energy to growth, reproduction, and hunting. The result has been catastrophic. With small mammals nearly wiped out in some areas, the Everglades' food web has been radically restructured. In this way, parasites are both gatekeeping host health and shaping whole ecosystems. When you remove them, invasive species can expand like an unchecked wildfire, leaving a trail of ecological collapse in their wake. The Burmese python is what happens when evolution loses one of its oldest regulatory tools.

As striking as the Burmese python is, its story illustrates what happens when ecological checks, like predators or parasites, are removed. But what about ecosystems where parasites are still in play, doing their quiet work of keeping things in balance? To really appreciate parasites as ecosystem engineers, we need to flip our assumptions. Far from being freeloaders, parasites can serve as keystone species—organisms whose presence and influence disproportionately shape the ecosystem they inhabit. Take the salt marshes of the American Pacific Northwest, for example. These brackish coastal systems are teeming with life: fish, shorebirds, snails, and worms crisscrossing a muddy, sea-scented web of interactions. And threading through this mess? Parasites. Lots of them. Among the most important are the trematodes that we encountered earlier. These trematodes link all of these salt marsh creatures together in intricate ways. Infected killifish, for instance, swim erratically, flashing their bellies in a way that attracts birds. This boosts transmission to the next host, but it also shifts predation patterns in the ecosystem. Birds eat more fish, fish populations decline, snail populations change, algae grow differently—and the whole system ripples in response.

Ecologists used to think of parasites as noise in the ecological signal. But a growing body of work, including network studies of food webs, has shown that when you map out who eats whom, and you include parasites, the web does not just get bigger, it gets more connected, more complex, and in many ways, more stable.[35] Parasites increase biodiversity by adding new species and links between species, especially across trophic levels. They create indirect effects, cascading consequences of infection, that would otherwise go unnoticed. So, when invasive species escape their parasites, as the Burmese python did, it is not just a lucky break for the invader. It is a structural shift. And when native parasites are wiped out, whether through habitat loss, pollution, or climate change, it is not just a loss of biodiversity. It is a weakening of the very threads that hold the food web together. In this way, parasites regulate populations and are responsible for shaping the very blueprint of ecosystems. We just did not see them for what they were. Now with the right tools and models, we are beginning to map the hidden scaffolding they provide.

PUTTING IT ALL TOGETHER

Parasites are evolutionary signals—sharp, unrelenting, and often invisible to the naked eye. Where once we might have dismissed them as mere noise, hitchhikers, or degenerates in the grand story of life, we now see them for what they are: potent drivers of diversity, shapers of food webs, and subtle architects of form and behavior.

From manipulating limbs to rewriting courtship, from undermining ecological stability to nudging species toward speciation, parasites are active engineers of evolutionary change, not passive passengers. They exploit every loophole in biology, bypass every safeguard, and leave nothing untouched. But they do not simply destroy. They refine. They pressure. They select. In doing so, they carve paths that shape not only individual bodies but entire lineages and ecosystems. For every host defense, there is a parasite counterstrike. And for every adaptation, a hidden struggle for survival plays out at the microscopic level. To view evolution through the eyes of parasites is to see a different kind of natural history—one driven as much by infection and resistance as by competition and cooperation. It reminds us that the evolutionary story, while written in muscles and skeletons, is also scripted in life cycles, immune responses, and parasitic pressures.

We have seen how parasites mold bodies and behaviors, but their influence goes even deeper. In the next chapter, we descend to the molecular battlefield, where the scars of ancient infections remain etched in our genomes, where arms races are fought in protein folds and regulatory genes, and where the past lives on, not in fossils but in the DNA of the survivors.

2

THE GENOMIC BATTLEFIELD

For he said unto him, Come out of the man, thou unclean spirit.
And he asked him, What is thy name?
And he answered, saying, My name is Legion: for we are many.

—Mark 5:8–9, King James Bible

In a remote village in the heart of the Democratic Republic of Congo, a quiet crisis unfolds. Days ago, a group of children, ranging in age from eight to fourteen, were clearing brush and helping their families prepare the land for planting. They were, by all accounts, healthy, vibrant, and full of mischief. Then the fevers came.

At first, it seemed like nothing out of the ordinary. Mild headaches, hot foreheads, some fatigue—typical childhood complaints in a region where illness is all too common. Parents brewed herbal remedies; some of the children recovered. But for a handful, the symptoms deepened. They became withdrawn, then irritable. They lashed out, cried without reason, or sat in silence, staring into space. Village elders chalked it up to moodiness or spiritual imbalance. Others suspected malaria. Antimalarial drugs were administered, but nothing changed. Soon, the children could no longer stay awake in school. They nodded off in the middle of class, or while walking, or while eating. One boy collapsed face-first into his lunch. Teachers stopped letting them attend. Their bodies grew weaker. One child began to convulse. His limbs shook uncontrollably, his eyes fluttered, and his breathing became erratic. Panic spread. But no one could say what was wrong. And without a proper diagnosis, time was running out.

This is not a scene from a dystopian novel or a forgotten past. It is the present-day reality of human African trypanosomiasis (HAT), better known as African sleeping sickness. It is caused by *Trypanosoma brucei* (subspecies *gambiense*), a parasitic protozoan transmitted by the bite of the tsetse fly, its vector, and it remains one of the deadliest neglected tropical diseases in Central Africa. Luckily, a combination of increased and sustained surveillance, treatment programs, and vector control efforts has dramatically reduced the case numbers, from tens of thousands in the 1990s and early 2000s to under a thousand reported cases annually from 2019 until at least 2023. Despite this encouraging news, forty to seventy million people in sub-Saharan African remain at risk.[1] Unlike acute infections that strike quickly and resolve fast, *T. brucei* is a slow invader. It hides in the bloodstream for months, sometimes years, slowly undermining the body from within. Its most insidious feature? It wears a shifting disguise. Like a master of costumes, *Trypanosoma brucei* covers its surface with a special protein "coat." Just as the immune system starts to recognize and attack, the parasite swaps that coat for a new one. These coats are made of what biologists call variant surface glycoproteins, or VSGs. Think of them like different outfits. The immune system might be able to recognize the red outfit, but by the time it does, the parasite is already wearing blue—then green, then yellow, and so on. This constant shape-shifting confuses the immune system and buys the parasite more time to survive and spread. This parasite plays a long game—one of attrition, manipulation, and, if left unchecked, death. And while its impact is devastating in the present, *Trypanosoma* has also left a mark in a much deeper, permanent place: our genome—the vast library that contains the blueprints for building and running our bodies.

What follows in this chapter is the story of not just a single parasite but an entire class of evolutionary antagonists—organisms that have shaped, scarred, and in some cases written themselves into the DNA of their hosts. Our genomes are not tidy blueprints but records of countless ancient infections. Some of the most consequential battles in evolutionary history have been fought not with swords or spears, but with strands of RNA, shifting proteins, and the endlessly adapting machinery of life itself.

THE GENOME IS A WAR ZONE

On April 25, 1953, a seminal letter was published in *Nature* titled "Molecular Structure of Nucleic Acids: A Structure for Deoxyribose Nucleic Acid," a one-page masterpiece by James Watson and Francis Crick.[2] In it, they unveiled the elegant double-helix model of DNA, based on Rosalind Franklin's iconic "Photo 51" X-ray diffraction image. That moment changed everything. Until then, biology had no clear explanation for how traits passed from parent to child. Watson and Crick's model did not just reveal a beautiful spiral; it revealed the mechanism of inheritance and replication. As they modestly put it: "It has not escaped our notice that the specific pairing we have postulated immediately suggest a possible copying mechanism for the genetic material."

The discovery was a whirlwind. Building on decades of groundwork, from Miescher's identification of DNA in the 1860s, Chargaff's rules in the 1940s, and Franklin's crystallography, their breakthrough unlocked the blueprint of life. It was the smoking gun that launched molecular biology into a new era. In the decades since, we have moved from reading genomes to editing—think CRISPR, gene therapies, ancient DNA sequencing, all ripples from that moment in 1953.

Every living cell contains DNA, written in a four-letter chemical code—A, T, C, and G—that acts like a biological blueprint. As Watson and Crick revealed in 1953, DNA's double-helix structure enables precise copying, allowing cells to pass down traits from generation to generation. Within this blueprint are genes—stretches of code that instruct the cell how to build proteins. These proteins drive every action we take, from digesting food to fighting infections. But DNA is not a static blueprint. It flows into action through RNA, the go-between molecule that reads DNA's instructions and helps assemble proteins. And these proteins shape us: our traits, our defenses, even our susceptibility to diseases. Random changes to this genetic code, called mutations—whether a single letter swap, deletion, or rearrangement—are the raw material of evolution. Most mutations are neutral or harmful, but occasional beneficial ones can help organisms resist disease, live longer, or reproduce more successfully. Natural selection then amplifies these useful variations.

Beyond changing the code itself, parasites and other environmental factors can tweak how our genes are used—without rewriting the DNA. This is the role of the epigenome, a system of chemical tags (like methyl groups, which is a carbon atom bonded to three hydrogen atoms) and molecular switches that tell cells when to turn on or off. The epigenome is fascinating. It adds an extra layer to genetics, reminding us that DNA is not destiny. Rather than being a rigid code, our genetic material is wrapped in a dynamic system of chemical "tags" that can switch genes on or off. These marks are responsive to diet, stress, environmental toxins—and yes, infections too. In essence, the epigenome helps explain why our genes do not tell the full story: They are constantly being tuned and reshaped by our experiences, not just inherited from our parents. For instance, some parasitic worms can reprogram host immune genes, changing how they respond to future infections.[3] In effect, they leave molecular "Post-it notes" on our DNA, reshaping gene activity.

So, the genome not only documents evolutionary history but also bears the molecular scars of ancient battles, through mutations and epigenetic marks left by parasitic pressure. This raises a critical question: How much of who we are, even down to our genes and their regulation, was sculpted from within by the organisms we fought?

MOLECULAR SCARS: PARASITES THAT WROTE THEMSELVES INTO OUR DNA

We like to think our genomes are ours alone, a biological heirloom passed down from our parents, shaped by the chance encounter that brought them together. That inheritance stretches further still: to our grandparents who crossed oceans in search of new lives, and back through time to hominid ancestors who once roamed the plains of Africa and Eurasia. But hidden within that intimate script are stowaways—genetic invaders whose blueprints are now interwoven with our own. These are the remnants of ancient viral infections, fossil traces of parasites that did not just infect us but became part of us. In this sense, we are both human and genomic mosaics—chimeras.

Viruses are the most intimate parasites I know, more so than any worm I have studied. And I study some truly bizarre ones: marine worms that bore into oyster shells, secrete acid, and raise broods that cannibalize one another in their egg sacs. But viruses? Let's put it this way: There are levels to this game called parasitism. A virus is essentially a fragment of genetic code, often wrapped in a protein shell, with no metabolism and no self-sustaining life, that infiltrates a host cell and hijacks its machinery to make copies of itself. What fascinates me most about these entities, teetering on the boundary between life and nonlife, are three traits that likely fueled their evolutionary success: (1) *simplicity*—many viruses are little more than bare strands of RNA or DNA; (2) *rapid evolution*—with every infection, they mutate, spawning within days a swarm of variants, or what biologists call quasi-species; and (3) *genomic hijacking*—some viruses splice themselves into the host's DNA, embedding like rogue sentences in a book's manuscript.

In terms of parasitic cunning, tapeworms and even tiny protozoans like *Trypanosoma* are brute forces; viruses are subtle infiltrators. And over evolutionary timescales, they may be even more consequential. Because they are acellular, viruses have left no fossil record. But most virologists believe they arose sometime after the last universal common ancestor (LUCA)—the ancestor of all cellular life—gave rise to bacteria, archaea, and eukaryotes like us. Since then, viruses have been slipping into genomes across the tree of life. Astonishingly, our own genetic code still bears their signature. In some cases, viruses reshape the organism they infect. Imagine a virus that does not just make you sick and move on but slips into your cells, edits your DNA, and then gets passed down to your children, and your children's children, like an unwanted family heirloom.

That's what happens with a strange class of viruses known as retroviruses. These viruses infect the body and aim straight for the blueprint of life. Retroviruses have a unique track. Unlike most viruses, which carry instructions in the form of DNA (like most life-forms on our planet), retroviruses use RNA to store their genetic information. Recall from the previous section that in almost all organisms, RNA works like a helper molecule, transcribing the information from DNA into a form that the cellular machinery can read and interpret to build the molecules in the cell that do the actual work—proteins. Here we have a virus that is actually using RNA as its information archive. Such a molecule is incompatible with DNA. We cannot connect sequences of DNA and RNA together; it is like connecting two different languages and expecting a translator (your cellular machinery) trained in only one to be able to decipher it. Instead, when these retroviruses or RNA viruses infect a host, they bring along an enzyme called reverse transcriptase, which essentially rewrites their RNA into DNA. The DNA is then smuggled into the host's own genome, hiding in plain sight. If the infected cell happens to be a sperm or egg cell, then the viral DNA does not just affect the individual; it becomes part of the species. It is passed on, generation after generation, like graffiti carved into the walls of a palace.

Over millions of years, these viral insertions have added up. Today, approximately 8 percent of the human genome is made up of these ancient viral fossils. Biologists call them endogenous retroviruses, or ERVs.[4] They no longer behave like active viruses, they cannot leap from person to person, but they are still there, hiding in the genomic background like faded text in an old manuscript. Make no mistake: these were once full-blown parasites. They invaded our cells, hijacked our replication machinery, and tried to spread themselves as widely as possible. And even now, long after most have lost the ability to function, they still shape our biology. In fact, part of the reason our immune system exists in its current form is to keep these viral stowaways in check.[5] If some ERVs wake up again, due to infection, stress, or genetic misfires, they can contribute to autoimmune diseases or even certain cancers. Our cells keep them under lock and key, silencing them with epigenetic tools like chemical tags and molecular guards. And they are not just in humans. ERVs are everywhere: in mice, sheep, whales, elephants, birds, and even reptiles. One study found that ERVs have invaded nearly every vertebrate genome examined to date, like a common watermark of ancient viral warfare.[6] Evolution, ever the recycler, has taken some of these old parasitic sequences and given them new purpose. A few have even been co-opted to help regulate immune responses, as if the cell said, "Fine, you're staying. Make yourself useful." It's a deeply ironic turn: the invaders we once battled have, in some cases, been drafted into our defenses.

How do we know these strange sequences are the remains of ancient viruses? When biologists compare DNA from different species, they sometimes find identical viral sequences in the same spots, like bookmarks left behind by a long-gone reader. That tells us the virus infected a common ancestor. And within those sequences, we can still recognize the viral calling cards, genetic instructions that resemble those in modern retroviruses, now broken or mutated beyond repair. One viral gene in particular has earned a strange kind of fame: *syncytin*. Originally part of a retrovirus that infected our distant ancestors tens of millions of years ago, this gene was never meant to serve a higher purpose. It encoded a fusion protein designed to help the virus merge with host cells. But somewhere along the evolutionary timeline, that viral trick was co-opted by mammals and turned to our advantage. Today, syncytin is essential for forming the human placenta. It helps fuse cells together to create the syncytiotrophoblast, a specialized layer that allows a fetus to draw nutrients from the mother without triggering an immune attack.[7] It might seem odd that pregnancy would trigger an immune response, but a fetus is technically a foreign invader to the mother's body. Half of its genetic material comes from the father, making it genetically distinct. Under normal circumstances, this kind of foreign tissue would provoke an immune response. But during pregnancy, the maternal immune system undergoes special adaptations to tolerate the fetus, a process sometimes referred to as immune privilege, and syncytin is a key part of this.

Incredibly, this is not just a human story. Different mammal lineages—primates, rodents, and others—each adopted different retroviral fusion genes independently for the same task. In other words, pregnancy as we know it depends on ancient viral infections. Without syncytin and its viral cousins, mammals might never have developed internal gestation at all. This is a powerful example of evolution's dark alchemy—how a genetic parasite, once a threat, became a cornerstone of life itself.

Not all viral sequences become essential parts of our biology. Many are left behind as pseudogenes—genetic relics that no longer function, the broken remains of viruses that once tried to infiltrate our DNA. Unlike syncytin, which evolution borrowed and rewrote into a tool for building life, these remnants are mutationally crippled, unable to express proteins or replicate. Genome-wide surveys have revealed thousands of such pseudogenes scattered throughout our chromosomes, not only from ancient viruses but also from parasitic genetic elements that once hijacked our genome's machinery for their own replication.[8] Biologists now view many pseudogenes as fossil evidence of failed invasions, stray sequences from retroelements and defunct viral fragments, long since neutralized by mutation or epigenetic silencing. They are not doing anything useful, but they are not gone either. They persist like ancient graffiti etched into our DNA, a molecular record of past biological skirmishes. Our genomes are not tidy archives; they are battle-scarred documents, scrawled over by invaders, some of whom managed to shape our biology, and others that history left behind.

THE IMMUNE SYSTEM: EVOLUTION'S STANDING ARMY

The baby arrived listless and pale. Just six months old, her eyes were glassy, her skin warm with fever. She had stopped nursing the day before. Her mother, who had already lost one child the previous year, walked three miles through the brush of rural Angola to reach the nearest clinic. A nurse pricked the baby's heel, and a drop of blood revealed the culprit almost immediately under the microscope:

Plasmodium falciparum—the deadliest malaria parasite. But it was not the infection alone that caught the doctor's eye. The baby's red blood cells looked . . . odd. Many were misshapen, sickled, like crescent moons. What the doctor saw in her blood was evolution in action. Her sickle-shaped cells were the result of a mutation passed down for generations, forged in humanity's long war with malaria. That twisted hemoglobin, the same mutation that causes sickle cell disease, was also the reason she had not died yet. Her immune system, her very blood, bore the fingerprints of parasitic selection.

If our genomes are battlefields, then the immune system is the front line, a standing army trained to recognize, repel, and remember invaders. But this army did not spring into existence fully formed; it was forged over eons in the crucible of parasitic pressure. Every toll-like receptor, every signaling cascade, every fever response is part of the ancient molecular arms race between hosts and parasites, a Red Queen dynamic at the genetic level, where each side must evolve just to keep pace with the other. One of the clearest examples of this is the evolution of major histocompatibility complex (MHC) genes, the genetic instructions for molecules that present pieces of pathogens to immune cells. MHC loci are among the most diverse regions in the human genome. They evolve rapidly and exhibit extraordinary variation and diversity across individuals and populations.[9] Why? Because a parasite that cannot be seen cannot be stopped. If a pathogen evolves a way to avoid recognition, the host must evolve a new MHC variant to detect it. And so the spiral continues—a coevolutionary tug-of-war written in chemical letters.

This molecular arms race is especially visible in regions of the genome under positive selection, where beneficial mutations spread rapidly because they confer a survival advantage against infection. Nowhere is this clearer than in the genetic landscape shaped by malaria—a blood disease caused by a parasite that has stalked humans for tens of thousands of years, halting conquest, progress, and exploration throughout history (I discuss the parasite that causes malaria in more depth in chapter 7.) In parts of Africa, a single-letter mutation in the beta-globin gene causes red blood cells to deform into the infamous sickle shape. These cells struggle to carry oxygen efficiently and have difficulty passing through small blood vessels, leading to painful blockages and tissue damage. While sickle cell disease is debilitating in its homozygous form, carriers of one copy of the mutation (heterozygotes) are granted protection against the *Plasmodium* parasite. It is a cruel compromise: a defense mechanism that saves many, but harms some. This is not the only genetic shield honed by malaria's pressure. Mutations in G6PD, a gene involved in red blood cell metabolism, also confer protection against infection, but come at a cost. Individuals with certain G6PD variants are more susceptible to hemolytic anemia when exposed to specific foods or drugs.[10] These trade-offs underscore a brutal truth: Evolution does not engineer perfect solutions; it cobbles together workable ones that simply improve the odds of survival, generation by generation. As the economist Thomas Sowell once put it in his 2007 book *A Conflict of Visions*, "There are no solutions, only trade-offs"—a principle that resonates not just in markets but across the entire natural world.

Beyond malaria, parasites have sculpted our immune systems in subtler but equally profound ways. Toll-like receptors (TLRs), a family of proteins that detect microbial signatures, show signs of accelerated evolution in primates.[11] Some variants improve recognition of bacterial or parasitic molecules, while others reflect adaptations to ancient infections no longer circulating. What we inherit is not a universal template for immunity, but a patchwork of localized strategies shaped by the specific pathogens our ancestors faced. Immunity, in this light, is not a cooperative endeavor between host and pathogen. It is reactive. It is personal. It is the cellular equivalent of trench warfare: costly, entrenched, and deeply informed by past assaults. Every molecule we deploy in defense was selected not because it created balance, but because it helped us survive.

PARASITES AND HUMAN EVOLUTION: ARE WE THE PRODUCT OF INFECTION?

A strange idea has taken hold in evolutionary biology—one that asks whether our very humanity, from the genes that protect us to the behaviors that define us, has been sculpted by the constant presence of parasites.

Across generations, populations living in parasite-rich environments have developed distinctive genetic signatures. Take, for example, the populations living in areas where schistosomiasis—a parasitic flatworm disease spread through freshwater snails—is endemic. Studies have shown that these populations often exhibit selection in immune-related genes, including those influencing cytokine responses and T-helper cell balance.[12] In such regions, the immune system has evolved under intense pressure, forced to strike a delicate balance between eliminating parasites and avoiding self-destruction through autoimmunity.

The evolutionary tug-of-war is inscribed in our very DNA. The MHC genes, those immune recognition genes mentioned earlier, are among the most variable in the human genome. This hyperdiversity is no evolutionary accident. It is a signature of countless generations locked in battle with shape-shifting parasites. In regions with heavy parasite exposure, individuals with rare or novel MHC variants have an edge: Their immune systems can recognize threats others cannot. It is a genomic arms race, and the parasites are always just a step behind, gaining ground. But immune evolution does not stop at the blood or bone marrow; it is written in our skin and our guts.[13] When humans shed their fur, we also shed a layer of protection. Exposed and vulnerable, our skin had to evolve into more than a barrier: it became an immune organ in its own right. Layers of keratinocytes and sentinel immune cells now patrol the frontier between body and world, guarding against the microbial parasites that share our environment.[14] Even the presence of antimicrobial peptides like CIDAMPs—ancient molecules that date back to the origin of multicellular life—underscores how deeply embedded this defense system is.

Our guts tell a similar story. As detailed in recent studies, helminth worms trigger immune response but also shape the gut microbiota, which in turn feeds back into host immunity. In fact, coevolution with intestinal parasites may have helped regulate inflammatory responses in the gut, helping our ancestors avoid overreaction to harmless microbes.[15] The rise of autoimmune disorders in industrialized nations, places where parasites have largely vanished, has led to the "hygiene hypothesis": the idea that our immune system, shaped by millennia of parasitic exposure, now overreacts in the absence of those ancient foes.[16]

From our skin to our gut, our biology is not merely shaped by infection, it is in part the product of it. Evolution, it seems, is not a clean war against invaders. It is a long-term negotiation—sometimes brutal, occasionally collaborative, always transformative.

PUTTING IT ALL TOGETHER: OUR GENOMES REMEMBER

Our DNA is both a record of ancestral lineages and a battlefield log, inscribed with the scars, strategies, and compromises of millions of years of evolutionary conflict. Parasites have acted as powerful architects of genomes. They have dictated the terms of our biology, influenced the evolution of our immune systems, and in some cases embedded their genetic blueprints directly into ours. The syncytin gene that allows pregnancy, the pseudogenes that litter our chromosomes, and the immune receptors fine-tuned by protozoan assaults all whisper the same story: that survival has never been about purity but about adaptation.

These infections have left more than scars. They have left instructions—genetic residues and epigenetic signals that guide how our bodies respond to future threats. Our genomes are not passive archives. They remember; they adapt; they tell the story of a species in constant negotiation with its invaders. Without parasites, we might not have the same immune complexity, the same genetic variation, or perhaps even the same biological capacity for gestation and survival.

And the story does not end there.

In the next chapter, I turn to one of evolution's strangest tricks: the transfer of genes not just down family lines, but across species boundaries. Horizontal gene transfer—the great genetic heist—has allowed organisms to acquire traits overnight that would normally take generations to evolve. Once again, parasites, viruses, and microbial hitchhikers are at the center of it all.

3

THE GREAT GENETIC HEIST

I have this strange feeling that I'm not myself anymore. It's hard to put into words but I guess it's like I was fast asleep and someone came, disassembled me, and hurriedly put me back together again. That sort of feeling.

—Haruki Murakami, *Sputnik Sweetheart*

As a scientist, there are very few emotions that can rival the ones associated with discovery, especially the discovery of something truly novel. We observe nature, investigate phenomena, and comb through prior studies that have shaped our current understanding. And then a hunch surfaces. A small voice whispers: "Wait—there's something more to this. Something everyone else might have missed."

So we hypothesize. We design experiments. And whether you are a principal investigator at a research-intensive university commanding an army of postdocs or someone like me, at a small teaching college with a few brilliant, bushy-tailed undergrads, you dig in. You spend months, years, collecting data, troubleshooting equipment, and chasing patterns. Then suddenly, something unexpected appears—a result that does not quite fit. You repeat the experiment, double-check your methods, and obsess over the data. Slowly, you begin to realize you are seeing something new, something that had been hiding in plain sight. In those moments, adrenaline, dopamine, and serotonin all fire at once. You are not just solving a puzzle; you are revealing a hidden piece of nature's architecture. It is exhilarating, humbling, and completely addicting.

I often imagine this must have been how Charles Darwin felt when he returned to England after his voyage aboard the *Beagle*, his journals brimming with clues. Among the many scribbled observations and musings, one of the most iconic is found in his notebook from 1837, beneath a sketch of branching lines, what we now recognize as the first known depiction of a phylogenetic tree. Next to it, he simply wrote: "I think."

That crude little diagram would go on to inspire one of the most powerful concepts in biology: that all life on Earth shares common ancestry. From this simple branching idea, the entire field of phylogenetics was born. The word *phylogeny* comes from the Greek *phylon*, meaning "tribe" or "kind," and *genesis*, meaning "origin" or "birth." In essence, a phylogeny is a record of how living things are related to one another—a family tree for life on Earth. For more than a century, biologists have used these "trees" to trace how species split and diverge over evolutionary time.

To picture one without seeing it, imagine drawing a single vertical line—this is the trunk, representing a distant ancestor. Partway up, the line forks into two branches. Call those branches *Species B* and *Species C*. Because they split from the same branch point, they are each other's closest relatives; in evolutionary terms, they are *sister taxa*. Now imagine that lower on the trunk, before the branch that led to *B* and *C*, there is another fork leading to *Species A*. *A* sits on its own branch—it shared a common ancestor with *B* and *C*, but that ancestor lived further back in time. Real species follow the same patterns. Humans and chimpanzees, for example, are like *B* and *C*: They share a more recent common ancestor with each other than with any other living primate. This does not mean that humans "came from" chimpanzees. Instead, both species are like adjacent branches on the same tree—two twigs that sprouted from a shared ancestral branch millions of years ago. That is a massive distinction, and one that upended centuries of misunderstanding. This evolutionary framework marked a major paradigm shift from the older Linnaean system, named after the father of taxonomy, Carolus Linnaeus. The Linnaean system grouped organisms based on observable similarities: how they looked, functioned, or behaved. Phylogenetics still considers those traits, but it puts them in evolutionary context: Do these similarities reflect shared ancestry? That is the guiding question.

To build these trees, scientists gather massive amounts of data. This can include observable features like anatomy and behavior, or molecular data like DNA and protein sequences. Over nearly two centuries, evolutionary biologists have used these tools to trace the descent of genes, species, and entire lineages across deep time. But nature, as it turns out, is rarely that tidy. The more we studied genomes, especially with the arrival of molecular tools, the more cracks began to appear in the neat, treelike structure of inheritance. In some places, genes appeared to have jumped from one lineage to another. Whole stretches of DNA popped up in places they should not be. Instead of branches that only diverged, we began to see signs of convergence, insertion, and trade. It was as if someone had sneaked into the tree of life and rewired it from within. And so we arrived at one of evolution's most radical concepts: horizontal gene transfer (HGT). HGT is a process by which genes move sideways or laterally, across species boundaries—spreading not through reproduction, but through infection, absorption, or outright theft. Horizontal gene transfer reveals that evolution does not always proceed through slow modification of inherited genes. Sometimes entirely new genetic capabilities arrive all at once, acquired from another organism.

To say that HGT complicates phylogenetics is an understatement. It complicates it a lot. The theoretical framework of phylogenetics is based on ancestry—the idea that the genetic material we inherit is passed down vertically from parent to child, and from that child to their offspring, across generations and millions of years. This model assumes clean lines of descent, Darwin's branching tree, stretching from common ancestors into the diversity of life we see today. But horizontal gene transfer does not follow those rules. It scrambles them. In some cases, genes jump from one organism to another across not just species but entire kingdoms of life.[1] These lateral movements stitch together lineages that should have remained separate, turning the tidy tree of life into something far messier—a tangled web of genetic exchange, more fungal mycelium than oak.

Who are the main culprits behind this genomic smuggling operation? They are parasites, viruses, and microbial symbionts—life's hitchhikers, saboteurs, and opportunists. These organisms, often dismissed as biological nuisances, turn out to be agents of innovation, ferrying DNA across boundaries, sometimes hijacking it for their own gain, sometimes delivering it like a Trojan horse into the genomes of their unsuspecting hosts.

This raises some unsettling questions.

In AP biology textbooks, students learn early on that DNA's chemical structure makes it remarkably stable, a likely reason it became life's archival molecule, rather than its more fragile sibling, RNA. But again and again, we are reminded that in biology, context is everything. Molecules may behave one way in isolation—in a petri dish, under controlled conditions—but very differently when they are part of the grand, messy orchestra of a living cell. So, is the genome really a stable, inherited blueprint? Or is it a fluid, mosaic document, patched together over evolutionary time through invasion, infection, and cohabitation? Did life's complexity evolve purely through gradual invention, or was it accelerated by genetic acquisition, through life stealing from life?

In this chapter, I explore a hidden engine of evolution, one driven not by gradual inheritance but by theft. Parasites are not just rule breakers in biology; they are rule makers. From bacterial genes hijacked to power parasite metabolism to viral gene swaps that blur the boundaries between species, we will follow how horizontal gene transfer has rewritten the story of life. Our journey will take us through a rogues' gallery of microbial and multicellular parasites, each engaging in genetic larceny. We will see how *Leishmania*, the protozoan behind horrific human disfigurements, acquired bacterial genes that help it survive the hostile terrain of the human body. We will examine evidence that the ancestor of a parasitic worm may have stolen genes from bacteria, enabling its descendants to breach human cells. And we will uncover how the scourge of river blindness, caused by a nematode worm, owes part of its success to genetic theft. These biological heists point to a deeper evolutionary truth: Much of life's complexity was not invented from scratch; it was borrowed, stolen, and repurposed.

BACTERIAL ARMS RACES: THE FIRST GENETIC THIEVES

Long before horizontal gene transfer became a hot topic in evolutionary biology, we were already using it without realizing it. In 2019, *CBS News* reported a story that struck a chord with millions of Americans. A twenty-six-year-old man named Alec Smith died because he could not afford his medication. Alec had Type I diabetes and relied on daily insulin injections to survive. But when he turned twenty-six, he aged out of his parents' health insurance. The cost of his insulin, $1,300 per month, was more than he could manage. In an attempt to stretch his supply, Alec began rationing doses. He was later found alone in his apartment, having slipped into a diabetic coma.[2] For many outside the United States, this might seem surreal, even neo-draconian, but for uninsured American diabetics, the absurd price of insulin is a well-known reality. Let me be absolutely clear: The cost of producing insulin is low. Since the 1980s, we have used genetically engineered bacteria to produce human insulin through a process known as recombinant DNA technology. The method is well-established, efficient, and inexpensive to manufacture at scale. Yet insulin's retail cost has ballooned—not because of science, but because of politics, patents, and pharmaceutical profiteering. In the United States, it has become a symbol of medical innovation hijacked by market forces.

That irony is hard to ignore: a lifesaving drug, made possible by splicing a human gene into the bacteria *Escherichia coli*, is now caught in a web of human-made barriers. Yet, in a strange way, that synthetic achievement mirrors what nature has been doing all along. We did not invent the idea of moving genes between species. Bacteria did—billions of years ago. And they still do it today, with stunning speed and precision. Let us look at how bacteria engage in their own form of genetic exchange, and how this ancient microbial innovation gave rise to one of the most urgent medical challenges of our time: antibiotic resistance.

Long before genetic engineering and modification, microbes were already swapping DNA like baseball cards in a schoolyard. Horizontal gene transfer in bacteria can occur through three main pathways: (1) conjugation, (2) transformation, and (3) transduction. Bacteria can use any of these three or combinations thereof to swap genetic material.

Conjugation is the microbial version of direct file sharing. Imagine two bacteria pulling up next to each other like docking spacecraft. One of them—the donor—extends a tiny tubelike structure called a pilus (pronounced *pie-luss*), which acts as a bridge or straw connecting the two bacterial cells. Through this microscopic bridge, it sends over a small loop of DNA known as a plasmid. Plasmids are like genetic USB drives—portable, plug-and-play pieces of DNA that carry useful traits. Unlike the main bacterial chromosome, plasmids are not essential for survival, but they often contain bonus features like antibiotic resistance genes. Once the plasmid arrives in the recipient bacterium, it can be copied, shared again, or even integrated into that bacterium's own genetic code. Suddenly, a bacterium that was once vulnerable to antibiotics can now survive and spread the resistance onward. This kind of genetic handoff does not happen only within the same species. Bacteria from completely different lineages can exchange plasmids through conjugation, blurring the lines of microbial ancestry—i.e., you are not (*necessarily*) the father. In evolutionary terms, it is like borrowing pages from a neighbor's survival manual, no permission required.

Transformation is more of a microbial scavenger hunt. Some bacteria have the uncanny ability to pick up loose fragments of DNA from their surroundings, usually genetic scraps left behind by dead or damaged cells, and incorporate them into their own genome. This is a bit like finding a working smartphone on the sidewalk, recharging it, and using the apps as your own. This remarkable phenomenon was first observed in 1928 by a British microbiologist, Frederick Griffith, who noticed that harmless bacteria could become deadly when mixed with heat-killed virulent strains.[3] Something, some invisible substance, was passing from the dead to the living. It was not until 1944 that Oswald Avery, along with colleagues Colin McLeod and Maclyn McCarty, identified the culprit: DNA. Their discovery, that DNA could be taken up and used by other organisms, helped lay the foundation for modern genetics.[4] Decades later, scientists began to realize they could hijack this natural process for their own purposes. In the 1970s, Herbert Boyer and Stanley Cohen developed the tools to insert foreign DNA into bacteria using the plasmids we discussed earlier. This paved the way for the field of genetic engineering and soon led to one of the most important breakthroughs in medical biotechnology: synthetic human insulin. In 1978, the San Francisco–based biotech company Genentech used bacterial transformation to insert the human insulin gene into *E. coli.* Just four years later, the product—humulin—became the first FDA-approved genetically engineered drug.[5] For the first time, insulin was not being harvested from pigs or cows; it was grown inside bacteria. This is still how we make insulin today. By inserting a carefully designed piece of human DNA into a bacterium, scientists can effectively "program" it to manufacture insulin, growth hormones, clotting factors, and other lifesaving molecules. The is the backbone of recombinant DNA technology, which revolutionized medicine and still powers much of the biotech industry today.

The third pathway for horizontal gene transfer in bacteria, transduction, brings us to a microscopic war that has been raging for billions of years, the oldest biological conflict on Earth: bacteria versus viruses. Transduction enlists viruses as accidental couriers of DNA. Specifically, it involves bacteriophages, a special class of viruses that infect bacteria. When a phage invades a bacterial cell, it hijacks the bacterium's machinery to make more copies of itself. But sometimes, during this chaotic replication process, the virus mistakenly scoops up pieces of the host's DNA along with its own. When it goes on to infect a new bacterium, it delivers this genetic stowaway, essentially smuggling DNA from one cell to another. The result is a kind of accidental genetic remix, in which bacteria receive random bits of DNA from their viral invaders. Most of the time, it is junk. But occasionally, the cargo includes genes that offer a survival edge, like resistance to antibiotics or toxins to outcompete neighboring microbes.

Transduction is messy, random, and opportunistic. But in the microbial world, even accidents can be engines of evolution. These exchanges represent strategic adaptations in the microbial world. One of the clearest and most alarming outcomes is antibiotic resistance. Bacteria can acquire entire suites of resistance genes in a single transfer event, transforming them from vulnerable strains into formidable "superbugs" overnight.

And the gene swap meet is happening all around us.

Ever heard of MRSA—*methicillin-resistant Staphylococcus aureus*? Of course you have. It's a doctor's (and patient's) worst nightmare. MRSA started out as a mostly harmless skin bacterium, but it became a global health threat after picking up a powerful piece of DNA from another microbe. That DNA carried a gene called mecA, which rewired MRSA's internal machinery to resist antibiotics like methicillin and penicillin. Suddenly, drugs that once wiped it out became useless. This tiny genetic upgrade did not come from slow mutation; it came from a quick and dirty gene swap, likely from a related species and delivered via a phage or a plasmid. With one exchange, MRSA gained a molecular shield, and began spreading in hospitals, prisons, and communities around the world.[6]

More recent research is pulling the curtain back even further, revealing the hidden world of HGT among bacteria. For example, in 2020, researchers used a technique called Hi-C sequencing to track the movement of antibiotic resistance genes across different species of bacteria inside the guts of hospitalized patients.[7] The results were wild: Resistance genes were not just sitting around—they were on the move, jumping between unrelated bacterial lineages. Especially in patients undergoing antibiotic therapy, the pressure cooker environment of the gut seemed to amplify these genetic transactions, accelerating the rise of multidrug-resistant strains. Unfortunately, this is not a fringe phenomenon; it is mainstream biology, operating under our noses, and often in response to our own interventions. In fact, antimicrobials themselves may stimulate HGT. Conjugation is one of the main ways resistance spreads, and it often kicks into high gear when antibiotics are present. In other words, the very drugs we use to kill bacteria can actually encourage them to swap survival tools. This is like setting fire to a building and watching the occupants toss fireproof gear to each other as they run.

And it is not just pathogens.

Environmental microbes, exposed to low levels of antibiotics from livestock operations, wastewater, and soil, also participate in this gene economy, contributing to what scientists call a "global resistome." The problem is so acute that the U.S. Federal Drug Administration (FDA) developed the National Antimicrobial Resistance Monitoring System (NARMS) to track antimicrobial resistance in foodborne and other intestinal bacteria.[8]

This microbial underworld undoubtedly complicates our efforts to control diseases. But it also rewrites our understanding of evolution. Rather than slow, vertical inheritance from parent to offspring, bacteria treat the genome like an open-source platform—modular, remixable, and constantly updated. In this remix culture, parasites and viruses are key players. As we will see in the next section, they are not just recipients of genes; they are active couriers and disruptors of inheritance itself.

VIRAL SABOTEURS AND GENETIC HIT MEN

We have already seen how bacteria trade genes like currency. But some of the most prolific gene couriers on Earth are not even alive, at least not technically. Viruses—those minimalist biological machines—do not just infect cells; they infect evolution.

For billions of years, viruses have been hijacking, copying, and smuggling genetic material across the boundaries of species, kingdoms, even domains of life. In doing so, they have become one of the most powerful, if accidental, forces in horizontal gene transfer. Some act like genetic burglars, stealing host DNA and inserting it elsewhere. Others function more like hit men for hire, delivering specific genes that alter a host's behavior, immunity, or survival in dramatic ways. We have seen hints of this before. In chapter 2, we met syncytin, the gene responsible for forming the mammalian placenta, now known to be of viral origin. It was once part of a retrovirus that fused with a distant ancestor's genome. What was once an infection became an innovation. But viral genetic influence goes far beyond mammals. In the microbial world, viruses, especially bacteriophages, are constant agents of genetic chaos. They do not just kill bacteria; they move genes around like smugglers crossing a border.

The COVID-19 pandemic was a global tragedy, one that reshaped daily life and exposed the fragility of even the most advanced societies. For me, it hit painfully close to home when a childhood friend, someone this book is partly dedicated to, passed away in 2022 from complications of the virus. What made SARS-CoV-2 especially alarming to scientists and public health officials was that it was novel. This was not just another seasonal flu or a familiar coronavirus strain; it was a new arrival on the global stage. That meant its behavior—specifically its transmissibility, its pathology, even its basic biology—was initially a mystery. Its origin remains one of the most fiercely debated topics in recent scientific history.[9] Was it a lab leak? Did it really originate in bats and jump to humans via a wet market in Wuhan, China? Or was it something more nefarious? From podcasts to politicians, everyone who is not a virologist seems to have an opinion. But what do the data actually show?

As it turns out, the genome of SARS-CoV-2 is itself a mosaic—a patchwork assembled from fragments of other viruses. Recent genomic studies have revealed that key portions of the virus's genome, particularly the spike protein and nucleocapsid genes, show evidence of recombination events, a form of horizontal gene transfer between viruses. In fact, comparative analysis suggests that SARS-CoV-2 likely arose when viruses from bats and pangolins swapped genetic material, essentially cutting and pasting chunks of each other's code. This is known in the literature as template switching.[10] Some researchers even believe that a third, still unknown virus may have contributed, making the final genome of Sars-CoV-2 a genetic patchwork from multiple sources.[11] So, while some search for conspiracy or intentional design, the reality is that nature does not need a lab to engineer pandemics. It just needs time, hosts, and the ancient machinery of horizontal gene transfer. That is perhaps the most sobering takeaway of all: The same evolutionary processes that gave bacteria antibiotic resistance and gave us the ability to manufacture insulin also gave rise to a global pandemic.

Since we used a destructive pathogen as a case example to start this section, let us stick with this theme. *Vibrio cholerae* is the infamous bacterium responsible for cholera. Its evolutionary tenacity is reflected in the fact that it has been responsible for seven major pandemics since the early nineteenth century—the first in 1817 and the most recent ones in Haiti (2010) and an ongoing outbreak in Yemen.[12] Altogether, this microbe is responsible for the deaths of millions of people spanning centuries. The pathology is simple yet effective: *V. cholera* secretes toxins that bind to the receptors on intestinal cells, triggering a series of events that ultimately cause massive secretion of chloride ions into the gut. Water then follows the ions down the established concentration gradient via osmosis, leading to profuse, watery diarrhea. The body rapidly loses fluids and electrolytes, causing severe dehydration. This is accompanied by vomiting, muscle cramps, and if untreated, organ failure and then death, which can occur within hours of infection.

Here is the kicker: *Vibrio* is not especially dangerous on its own. But once it acquires a gene for the cholera toxin via a bacteriophage, it becomes deadly. This is what microbiologists refer to as lysogenic conversion in action: The phage inserts its DNA into the bacterium, turning it into a toxin-producing machine.[13] Suddenly, a mild-mannered microbe becomes a global public health threat, not by evolving on its own, but by picking up a viral hitchhiker. This is a form of transduction, discussed earlier as one of the three different ways viruses engage in horizontal gene transfer.

In addition to meddling in bacteria, certain giant viruses typically found in amoebas carry hundreds to thousands of genes for DNA repair, protein folding, and even translation, functions we once thought were exclusive to cellular life. These giant viruses blur the line between virus and cell and between invader and innovator, even swapping genes with their hosts, becoming hubs for genetic exchange across the tree of life. Some scientists have even argued that viruses belong in their own domain of life, a contentious claim that deserves an entire book on its own.[14] Even stranger are virophages—tiny viruses that infect giant viruses. One example, known as Sputnik, parasitizes mimiviruses inside amoebas and, in doing so, can shuffle genes between virus and host.[15] These systems create what scientists call mobilomes, dynamic networks of mobile genetic elements that facilitate gene transfer in all directions. It is horizontal gene transfer writ large, operating beyond species lines and deep into viral-host ecosystems.

So, when we talk about evolution, it is no longer a simple tree; it is a tangled web—one in which viruses have been stitching together strands of DNA between wildly different forms of life for eons. But the gene-moving talents of viruses are not limited to accidents or ancient history. Some viruses actively transform their hosts into biological weapons. One of the most dramatic examples is found not in microbes or mammals but in insects. Take parasitoid wasps that lay their eggs inside caterpillars. On their own, wasp larvae would not survive; they would be attacked by the host's immune system before they could hatch. But these wasps carry a secret weapon: the polydnaviruses. These are viruses that the wasp picked up and added to its genome. Millions of years ago, an ancestral virus inserted itself into the genome of a parasitic wasp, and rather than being destroyed, it became part of the wasp's genetic legacy, passed down to every generation since. When the wasp injects its eggs into a host, it also injects viral particles that suppress the host's immune response, alter its physiology, and manipulate its behavior, all to give the wasp offspring a better shot at survival. I discuss the details of this amazing partnership in the next chapter.

On May 22, 1860, Charles Darwin wrote a letter to his friend Asa Gray that included the following line that would become famous: "I cannot persuade myself that a beneficent and omnipotent God would have designedly created the Ichneumonidæ with the express intention of their feeding within the living bodies of Caterpillars."[16] One can only imagine Darwin's reaction had he known that, beyond the caterpillar's gruesome fate, nature had also armed the wasp with viral allies—molecular saboteurs that silence the caterpillar's last line of defense, granted not by a Creator's mercy but by millions of years of innovation. Regardless of the theological implications, this is a perfect example of how HGT can create brand-new functions: evolution not by invention, but by appropriation.

Just as viruses have shaped evolution, they may also help save it. With antibiotic resistance on the rise, scientists are turning back to phage therapy, a technique first explored in the early twentieth century and used for decades in Eastern Europe. As a reminder, phages are viruses that infect and kill bacteria, and because they are highly specific, they can target dangerous pathogens without harming the rest of the microbiome. In 2016, phage therapy made headlines when it saved the life of a man dying from a multidrug-resistant infection in the United States, using custom phage cocktails engineered from sewage samples.[17] In this new microbial arms race, viruses may once again serve as our allies, reminding us that the line between friend and foe is often a matter of context.

Ultimately, viruses are more than just pathogens. They are genetic freelancers, capable of turning the evolutionary tide by smuggling genes, disabling defenses, or rewriting genomes. Their ability to move DNA across species and kingdoms challenges the idea of evolution as a slow, linear process. Instead, it becomes a dynamic network of interactions, where even the simplest entities—bits of protein-wrapped code—can play starring roles in the grand story of life. In the next section, I take a closer look at how horizontal gene transfer has powered the evolutionary machinery of our larger macroparasites, the main players of this book.

PARASITIC PARTNERSHIPS AND DNA EXCHANGE

Trichinella: The Muscle Hacker

When most people picture parasitic roundworms, they imagine something out of a horror film; long, writhing creatures lurking in the intestines, showing up uninvited during colonoscopies or emerging like alien invaders from beneath the skin. This is an image straight from the 1990 cult classic movie *Tremors*, one of my personal childhood favorites. But one species, *Trichinella spiralis*, redefines what a parasite can be. This worm does not just invade tissues. It goes one step further: It invades individual muscle cells, making it the largest known intracellular parasite of humans. And what it does once inside is nothing short of amazing. Like a virus in worm form, *Trichinella* hijacks the host cell's machinery and transforms it into something entirely new: a "nurse cell," custom-built to shelter and nourish the parasite. The takeover begins by halting apoptosis, the cellular version of a self-destruct button. Apoptosis, more dramatically but accurately known as "programmed cell death," is a tightly regulated process built into our biology. It is what shapes fingers and toes in the womb, ensures that old or damaged cells do not linger, and acts as a crucial safeguard against infection and cancer. Think of it as an internal code embedded in every cell, ready to be activated if things go wrong. This cellular program relies on a precise sequence of gene activations and molecular signals, like entering the correct passcode on a terminal to initiate a clean shutdown. But in the biological world, tightly coordinated and coupled processes are vulnerable to failure if any part of that sequence is disrupted. That is exactly what *Trichinella* takes advantage of. It interferes with the host cell's genetic programming, silencing key pro-apoptotic genes and rerouting survival pathways. In doing so, it flips the script, converting a cell programmed to sacrifice itself in the name of the greater good into a long-term, living host. It is as if the parasite reaches into the cell's command center and rewrites the instructions mid-process: "Don't die. You're useful to me alive."

Then the worm gets to work.

It shuts down the muscle cell's identity, silencing key genes like actin and myosin, the tools that allow muscles to contract. The cell stops behaving like a muscle and starts behaving like a factory. New blood vessels sprout around it, a process called angiogenesis, ensuring that a steady stream of oxygen and nutrients flows directly to the parasite's door.

The worm also rewires the cell's metabolism, nudging it toward a more glucose-rich, energy-intensive state, almost like a miniature cancer, but one controlled by a parasite. A thick collagen capsule forms around the infected cell, shielding it from immune attack. Inside, the cytoplasm rearranges into a matrix that cradles the worm like a bespoke sleeping bag. This is biological reprogramming at an extraordinary level. *Trichinella* doesn't just live inside the host; it rewrites what the host cell is. For a parasite with no brain, no voice, and no limbs, it executes a molecular coup that rivals anything we see in nature, or in science fiction.

The ability of the *Trichinella* worm to hijack its host's cellular machinery has intrigued scientists for decades, largely because its behavior mirrors that of microbes like viruses and bacteria. Its uncanny ability to manipulate gene expression, suppress immune responses, and reprogram a host cell from within has raised a deeper, more provocative question: Could this resemblance be rooted in something more cryptic than evolution? Traditionally, such similarity between unrelated organisms has been attributed to convergent evolution, the process by which different species independently evolve similar traits in response to comparable environmental pressures. But in the case of *Trichinella*, the sophistication of its molecular tool kit has prompted researchers to consider that horizontal gene transfer might have been the culprit that gave *Trichinella*'s ancestors its superpowers. Indeed, recent studies led by Dante Zarlenga of the Animal Parasitic Diseases Laboratory at the U.S. Department of Agriculture have provided tantalizing evidence for this. In one study, *Trichinella* was found to harbor a functional gene that codes for the enzyme cyanase, which breaks down toxic cyanate into harmless products. What makes this especially remarkable is that mammals, including *Trichinella* hosts, do not produce cyanate. Phylogenetic analyses suggest that this gene was acquired by a deep ancestor of *Trichinella* from a member of the plant kingdom or a symbiotic microbe associated with plants, making it one of the rare documented cases of eukaryote-to-eukaryote gene transfer in a parasitic nematode lineage.[18] The function of cyanase in *Trichinella* remains enigmatic, since its host does not typically accumulate cyanate. But scientists hypothesize that the peculiar environment of the nurse cell might foster buildup of metabolic byproducts like cyanate. In this context, the parasite's cyanase gene could serve as a biochemical lifeline, detoxifying waste or even harvesting it as an energy source.

This suggests something more profound: that the intimacy of parasitism—not just physical but genomic—can open doors to unexpected evolutionary transactions. Over millions of years, *Trichinella* may have not only co-opted the host's cellular machinery but fused its own molecular repertoire with genes pilfered from plants and microbes. Such gene acquisitions likely conferred adaptive advantages, helping the parasite survive within a metabolically unusual, immunologically fortified niche. In the ongoing arms race between host and parasite, a genetic heist like this could mean the difference between extinction and evolutionary success.

Leishmania: The Creative Smuggler

In the early years of the Syrian civil war (2011–2015), most eyes were drawn to the immediate devastation; bombed-out cities, streams of refugees, and the collapse of once-functional institutions. But another crisis was quietly unfolding beneath the surface, one that did not make headlines until the scars began to show, literally. A parasitic outbreak began to spread through refugee camps and war-torn regions, with outlets such as Al Jazeera providing heart-wrenching videos of children with scars and boils on their hands, feet, and faces, wincing at the repeated drug injections necessary to stem the progression of the disease. The affliction became locally known as "Aleppo boil."[19] In the clinical literature, it is called cutaneous leishmaniasis. What began as isolated lesions on the skin became an epidemic in slow motion, with tens of thousands of new infections spreading across Syria, Lebanon, Turkey, and Jordan. The culprit was not a worm, a bacterium, or a virus, but a protozoan: *Leishmania*, a single-celled eukaryote that infects human immune cells, specifically macrophages, and replicates from the inside out. Whereas *Trichinella* remains hidden from the immune system with the help of its hijacked muscle cell, now its own nurse cell, *Leishmania* actually infiltrates the immune system and uses its host's own defenses as a Trojan horse.

The life cycle begins with a bite. When an infected sand fly sinks its needlelike mouth into human skin, it delivers tiny, whip-tailed parasites into the wound. These swimming forms—think of them as the parasite's mobile scouts—immediately find themselves face to face with the body's immune defenders: macrophages, the cells whose job is to hunt and destroy invaders. But instead of being killed, *Leishmania* pulls off a biological con. It sheds its "tail," changes its shape, and settles comfortably inside the macrophage, like a spy slipping into enemy headquarters unnoticed. There, in the acidic environment meant to digest it, the parasite not only survives but thrives. It multiplies, eventually bursting the cell open and moving on to infect others, turning the body's defense system into a delivery service. When another sand fly comes along and bites the infected person, it slurps up these now-hidden parasites. Inside the fly's gut, they undergo yet another transformation, regrowing their tails and preparing for the next host. And just like that, the cycle starts all over again.

In the chaos of war, this cycle became amplified. Overcrowded camps, poor sanitation, and the collapse of vector control programs gave *Leishmania* an ecological jackpot. What had once been a relatively contained disease (accounts of leishmaniasis in the Middle East have been found in the ancient world as far back as 650 BCE) began to spread rapidly through vulnerable populations.[20] And the form it took in Syria was particularly cruel. Known as cutaneous leishmaniasis, this version of the disease does not kill—but it scars. After the bite, red papules erupt at the site of infection, slowly developing into ulcerating lesions that can take months, even years, to heal. These are not mere blemishes; they are open sores that disfigure the face, hands, and limbs, often in the most visible places. The psychological toll was devastating. For Syrians already dealing with trauma, displacement, and the collapse of public health systems, these wounds became yet another burden, a stigmatizing mark of both disease and war. Children were especially vulnerable. With no functioning infrastructure to manage the outbreak, and no access to consistent treatment, *Leishmania* spread with impunity, turning war-torn towns into perfect breeding grounds for both flies and fear.

As a neglected tropical disease, leishmaniasis has lagged far behind in terms of research into its control and management, compared with illnesses more common in wealthier Western nations. While billions have been poured into the development of new treatments for cancer, diabetes, and HIV/AIDS, *Leishmania*—a protozoan parasite that quietly afflicts some of the world's poorest populations—has received only a trickle of funding and attention.[21] The result? A handful of outdated, often toxic drugs that have barely evolved in decades. The first-line treatments for leishmaniasis were once compounds of antimony, a heavy metal with origins dating back to the early twentieth century. Drugs like sodium stibogluconate and meglumine antimoniate were widely used for decades, despite their serious side effects, lengthy injection regimens, and uncertain efficacy.[22] Over time, however, the parasite fought back. In places like Bihar, India—one of the hardest hit regions for visceral leishmaniasis, a form of the disease that results in a distended abdomen and spleen (common name: "Kala-azar")—*Leishmania*, specifically the species *Leishmania donavani,* developed widespread resistance to antimonials, rendering them virtually useless in some areas. This spurred the introduction of amphotericin B, a powerful antifungal that also happens to work against *Leishmania*. But this drug came with its own issues: It required intravenous infusion, came with a risk of kidney damage, and was prohibitively expensive in its most bioavailable form, limiting access in many endemic regions.[23]. Then came miltefosine, hailed as a breakthrough in the early 2000s. It was the first oral drug developed for leishmaniasis and sparked hope for a simpler, more accessible treatment option. Originally designed as an anticancer agent, miltefosine worked well in its early trials and quickly became a frontline therapy, especially in South Asia.[24]

But that optimism did not last. Within a decade, cases of miltefosine resistance began emerging, particularly in Bihar again, where incomplete treatment courses, improper dosing, and a lack of follow-up allowed the parasite to adapt. By the late 2010s, treatment failures were being reported at alarming rates, and resistance was no longer an isolated issue.[25] Even combination therapies, using miltefosine alongside amphotericin B or paromomycin, have shown signs of waning efficiency in some regions. Part of the problem lies in *Leishmania*'s genomic plasticity: The parasite can amplify certain genes, rearrange chromosomes, and, as recent research now suggests, exchange critical genetic material through horizontal gene transfer.[26]

New research is revealing that a not-so-small part of *Leishmania*'s genome may have been stolen from bacteria. One systematic survey from as far back as 2007 of the *Leishmania major* genome uncovered more than forty genes that appear to have bacterial origin, not shared with the parasite's related eukaryotic cousins.[27] Many of these acquired genes encode metabolic or enzymatic functions, particularly in sugar and amino acid processing, suggesting that they provide novel biochemical tools not previously part of the parasite's tool kit. Importantly, among the *Leishmania* gene transfers, ten of about twenty tested genes are conserved across multiple species, indicating that some bacterial genes were retained and potentially adapted for core functions in the parasite's biology.[28] In other words, *Leishmania* is not just shaped by its lineage. Deep in its genome lies a mosaic of bacterial DNA, snatched from microbes likely encountered in its insect niches, then retained and repurposed. These transferred genes may well have contributed to *Leishmania*'s ability to survive under stress, metabolize odd substrates, resist drugs, or invade hosts more effectively.

Recall that earlier we discussed the three classical pathways through which organisms exchange genetic material: transformation, transduction, and conjugation. Each of these originated in bacteria and involves different methods of DNA transfer. *Leishmania*, however, takes a decidedly different route. Instead of relying on these bacterial pathways, this parasite has evolved a unique genetic shuttle system known as extracellular vesicles, or EVs. EVs are common at the cellular level in all organisms, and they are often used to shuttle molecules like proteins and chemical tags from cell to cell in the body, but *Leishmania* uses it for more sinister purposes. Think of EVs as tiny, robust cargo ships, small packages released from one cell into the surrounding environment. These tiny vessels are sturdy enough to safely carry DNA fragments, shielding them from destruction until they dock seamlessly onto another cell. Upon docking, these vesicles deliver their genetic cargo directly into the new host cell, effectively "uploading" the genes. Recent groundbreaking research has revealed that *Leishmania* uses these vesicles to exchange critical survival genes, including those conferring drug resistance.[29] EVs released by drug-resistant *Leishmania* parasites can actually alter the drug sensitivity of previously susceptible parasites. In other words, EVs serve as a lifeline, rapidly spreading resistance genes through parasite populations, both within individual hosts and potentially across different *Leishmania* species. This EV-mediated mechanism, aptly termed vesiduction, provides parasites with rapid adaptive capabilities. By facilitating gene exchange without the direct contact of conjugation, and without the vulnerability of free-floating DNA involved in transformation, EVs offer a highly efficient and protective means for horizontal gene transfer.

What makes this particularly concerning from a clinical perspective is how quickly EV-based gene transfer allows *Leishmania* populations to adapt to hostile environments, such as those encountered during drug treatment. For patients, this means that drugs like miltefosine and amphotericin B, previously reliable treatments, are increasingly ineffective. Interestingly, we are now recognizing that by targeting the EV pathways and disrupting vesicle formation or release, it may become possible to block or slow down this alarming spread of drug resistance. However, this avenue for treatment remains theoretical to date; more funding into the basic research associated with HGT in this species will be needed before any real therapeutic treatment can be discovered.

Onchocerca and Wolbachia: The Parasitic Chimera

It begins with a fly.

You're walking along a riverbank, the sun high and unforgiving, the afternoon thick with humidity and the quiet hum of life. The cassava fields behind you are nearly cleared, and your woven basket of roots sits heavy at your hip. You pause to wipe your brow. That is when you feel it: just a pinprick, just a tickle. A black fly, small and unremarkable, lands on your skin. It bites. You slap at it, annoyed. You do not know it yet, but it has already done what it came to do.

In that fleeting moment, dozens of microscopic larvae of the parasitic nematode worm *Onchocerca volvulus* slipped beneath your skin through the fly's proboscis. You will not notice a thing for weeks. Maybe months. Not at first. These parasites are one of the quiet invaders.

The first sign is subtle: an itch that will not go away. It starts on your back or your arms, a maddening crawl beneath your skin. You scratch until it bleeds. The villagers call it "craw-craw." It is not the kind of itch that heals; it worsens, spreads, consumes. Beneath the surface, the larvae are growing, coiling into slender adult worms, and nesting in knots of flesh called nodules. They are mating now. The female can live up to fifteen years, producing tens of thousands of microfilariae offspring that pour into your lymph and blood, spreading to every corner of your body.

Your skin thickens. It loosens and wrinkles, looking decades older than you are. Small bumps rise beneath the surface, then harden.

One forms near your hip, another near your shoulder. Inside each, the worms are coiled like thread. You do not know they are there. Not yet.

Then comes the eye. The microfilariae drift into the aqueous humor, the clear fluid in the front of your eye. You see floaters at first, like dust or threads drifting across your vision. Then the inflammation starts. Your eye becomes red, sore, and light-sensitive. You begin to squint, even at dusk. The vision in your left eye becomes hazy, like a film drawn over it. You rub it. Nothing clears. The other eye soon follows. The immune system, in its fury, attacks the cornea, inflaming it, scarring it. The sclerosing keratitis begins—a whitening of the eye's surface that turns light into a blur, color into smudge. There are still shadows, movement, but no clarity. The world becomes a watercolor left in the rain.

Over time, the optic nerve suffers too. Its fibers shrink and die. The damage is irreversible. Your world narrows. The faces of your children lose their shape. You sit more often, hands outstretched, feeling for the doorway. You no longer go to the river. You no longer walk alone.

And one day, without ceremony, the last flicker fades. The darkness becomes whole.

You were bitten once, years ago, by a fly you barely noticed. Now the world is gone. This is the penultimate pathology of *Onchocerca volvulus*—a parasitic nematode that uses humans as its definitive host and relies on the bite of a blood-feeding black fly (various species of the fly genus *Simulium*) for transmission. These flies breed in fast-flowing rivers and streams across sub-Saharan Africa, making villagers who spend extended time near these waters, washing clothes, fishing, or performing daily chores, especially vulnerable. Hence the name: river blindness. The eventual loss of vision does not come directly from the worms themselves, but rather from what happens when they die, especially when microfilariae migrate into ocular tissues. Upon their death, the worms' bacterial endosymbiont, *Wolbachia*, is released, triggering a powerful immune response. It is this inflammatory cascade, aimed at the bacterium, that damages the delicate structures of the eye, leading to progressive vision loss and, in many cases, complete blindness.

The story of *Onchocerca volvulus* and its microbial partner in crime, *Wolbachia*, is of particular interest to us. As we will see in chapter 6, most symbiotic relationships maintain a clear boundary between organisms. Even in the most intimate partnerships, like coral and algae or fungi and photosynthetic bacteria, each partner retains its own genome, its own identity. But sometimes that line blurs. In *Onchocerca* and *Wolbachia*, evolution seems to have taken scissors to that boundary and begun stitching the two genomes together. If *Leishmania*'s use of vesicles to smuggle genes is clever, and *Trichinella*'s potential eukaryote-to-eukaryote gene theft is rare, then *Onchocerca* and *Wolbachia* represent HGT in its most extreme manifestation—a parasite made more parasitic by a bacterium whose genes now live on inside it. This is no simple infection, nor is it merely symbiosis. It is a genomic merger. Over the past two decades, genomic studies have revealed that *Onchocerca volvulus* harbors extensive stretches of *Wolbachia* DNA embedded in its nuclear genome. Some of these bacterial genes are no longer functional, like fossils of an ancient genomic invasion. But others are still active, transcribed into RNA, and potentially playing roles in the worm's metabolism, development, and interaction with its host. This is not just evidence of a past relationship; it is a sign that the worm's very biology has been reshaped by foreign genes.[30]

What is especially fascinating is the source: *Wolbachia* itself is an endosymbiont—a parasitic bacterium that became indispensable. In *Onchocerca*, *Wolbachia* is no passive resident. The worm cannot reproduce, develop properly, or even survive without its bacterial partner. And while other parasitic worms have lost their *Wolbachia* entirely over evolutionary time, *Onchocerca* has done the opposite. It has internalized it—physically, functionally, and genetically.

Horizontal gene transfer from *Wolbachia* into its nematode host likely occurred in ancient, ancestral worms, before the divergence of modern filarial lineages. The inserted genes appear scattered throughout the worm genome, including in intergenic regions and even inside introns of native worm genes. It is believed that some of these insertions could code for proteins associated with special functions that could directly support *Onchocerca*'s survival in oxygen-deprived or immunologically hostile tissues. While the full functional relevance of these transferred genes is still being unraveled, their sheer presence suggests an intimate coevolutionary fusion, one in which the parasite has begun to incorporate its microbial symbiont into its own blueprint. In this sense, *Onchocerca* does not just carry a bacterium; it carries its legacy. Even if *Wolbachia* were someday lost, its molecular fingerprints would remain inside the worm's DNA, like a ghost in the machine. This makes the *Onchocerca–Wolbachia* relationship not just a case of symbiosis, but a case of what some biologists call symbiogenesis: a merger of genomes that parallels the ancient endosymbiotic events that gave rise to mitochondria and plastids.[31]

In this molecular heist, the lines between parasite and partner, self and other, have all but vanished. What results is not two organisms working together but something entirely new: a genomic chimera, built on borrowed code.

GENETIC HIJACKING IN THE ANTHROPOCENE

We have spent this chapter following nature's most cunning genetic burglars: parasites, viruses, and symbiotic hitchhikers who break the rules of inheritance. But now the tables have turned. The thief has become the biotechnologist. We have entered a new evolutionary era, one in which humans are now the hijackers, using the same molecular tricks once employed by microbes and parasites to reprogram life. Welcome to the Anthropocene, the age in which nature no longer holds a monopoly on horizontal gene transfer.

The story begins with bacteria.

Faced with viral invaders, many bacteria evolved an ingenious defense system: They cut the viral DNA and store bits of it in their own genome like molecular mugshots. These stored sequences become part of the CRISPR system, a primitive immune memory. When the same virus attacks again, the bacteria recognize the sequence and use a protein (Cas9) to slice DNA apart, neutralizing the threat. In 2012, a small cadre of scientists, including Jennifer Doudna and Emmanuelle Charpentier, realized that this microbial system could be retooled as a programmable gene-editing tool, for which they jointly won the Nobel Prize in chemistry in 2020.[32] For those interested in the fascinating story of CRISPR and the scientists behind it, I strongly recommend Walter Isaacson's *The Code Breaker*. CRISPR-Cas9 gave us, for the first time, a way to cut and paste genes with the precision of a scalpel.

CRISPR does not just edit genes; it rewrites the rules of heredity—deliberately, predictably, and at scale. We can now insert jellyfish genes into mice, knock out inherited diseases in embryos, or resurrect ancient traits in endangered species. More recently, the biotech company Colossal Biosciences used CRISPR to make twenty precise edits in a gray wolf to modify its genome to match that of the long-extinct dire wolf, a small step toward the possibility of de-extinction technology.[33] HGT, once the domain of viruses and parasitic worms, has become the technological extension of human will.

Here is the twist: We did not just discover gene editing; we stole it.

CRISPR came from bacteria. Viral vectors, the delivery trucks of modern gene therapy, come from repurposed retroviruses. Even transposons, the "jumping genes" now used in genetic engineering, were first identified by Barbara McClintock in corn, but they are present in countless parasites. Many of the tools now powering synthetic biology were taken from the natural playbook of parasites. We have rebranded viral infection as "gene therapy" and parasitic DNA as "transgenic technology." The molecular saboteurs we once feared have become our lab partners. This is the great irony of the Anthropocene: We are now using parasites to become parasites, embedding our edits into genomes that never asked for them.

Perhaps the most chilling example of this is the gene drive. Normally, an altered gene has only a 50-50 chance of being passed down. But a gene drive rigs the system, ensuring that the edited gene is inherited nearly 100 percent of the time. CRISPR-based gene drives work by inserting an edit into one chromosome and then using CRISPR to copy that edit into its unedited counterpart. The result is near-guaranteed inheritance. It is not just a genetic tweak; it is a molecular coup. Imagine a mosquito population edited so that every female offspring is sterile. Release just a few engineered individuals, and in a few generations the entire population collapses—no smoke, no poison, no mass spraying, just genetic silence.

This is not hypothetical. It is already happening.

In Burkina Faso, researchers from the nonprofit consortium Target Malaria have staged the early chapters of this biological operation.[34] Their goal is to eliminate *Anopheles gambiae*, the primary mosquito vector of malaria in sub-Saharan Africa. Laboratory trials have already demonstrated the staggering potential. Within eleven generations, entire populations collapsed under the weight of their own broken heredity. In Brazil, the biotech firm Oxitec took a slightly different approach with *Aedes aegypti*, the mosquito responsible for dengue and Zika.[35] They released genetically engineered males designed to pass along a gene that kills their offspring before adulthood. In some regions, populations dropped by more than

90 percent. These modified insects were not gene drives in the strict sense. But the principle was the same: program a parasite to self-destruct.

These are not just acts of applied science. They are stories of molecular puppetry, of humans turning nature's most subversive tools against itself. Gene drives are parasitic technologies, built to invade and dominate a genome. They hijack the same evolutionary principles that parasites have used for millennia. Only now, we are the ones pulling the strings. There is a strange symmetry here. The same mechanisms we once saw as violations of the natural order—viral insertions, parasitic DNA, horizontal gene transfers—are now tools in our bioengineering arsenal. What nature did haphazardly over billions of years, we now attempt with intention, and at speed.

This power comes with a question: Have we become what we once feared?

In *Frankenstein*, Shelley warned of creation without humility. In *Parasite Rex*, Carl Zimmer marveled at the audacity of parasites that could rewrite their host's biology. But now, it is we who rewire embryos, we who insert foreign DNA, we who deploy viruses to cure, or to control. The parasite once lurked at the edges of biology—an intruder, an invader, a cheat. But in the Anthropocene, the parasite has become a blueprint, and we are following it. In the final chapter, I return to this theme and discuss what the future may hold for us humans who are now leveraging the techniques of parasites for our own benefit and ends.

PUTTING IT ALL TOGETHER: THE GENOME AS AN ARCHIVE OF INTRUSION

We often speak of the genome as a blueprint, as if it were a tidy architectural plan handed down from parent to offspring, generation after generation. But the truth is far messier, and far more thrilling. As we have seen throughout this chapter, evolution has never been a closed system. It has always been porous, hackable, vulnerable to intrusion. Life did not evolve in isolation; it evolved amid theft, sabotage, and collaboration.

The genome is not sacred. It is a palimpsest, a manuscript overwritten time and again by outsiders: bacterial genes smuggled into parasites; viral relics buried in our own DNA; symbionts stitching their sequences into hosts. Even now, human hands are at the typewriter, inserting edits, building drives, rewriting futures. These invasions have shaped everything from metabolism to reproduction to the development of complex bodies. The genes that once entered by stealth have become foundational, co-opted, domesticated, and made indispensable. The very act of giving birth, for instance, depends on syncytin, a gene of viral origin. We owe our existence to a retroviral insertion that turned maternal tissues into something evolutionarily novel. In this way, parasites are architects of innovation. Their survival strategies have forced genomes to adapt, collaborate, and evolve in unpredictable directions. Their intrusions have rewired evolutionary trajectories and, in so doing, have helped shape the living world as we know it.

If we take a step back and look at the genome not as a clean lineage but as a living archive, we begin to see the fingerprints of these biological thieves everywhere. Introns, transposons, viral insertions, horizontally transferred genes, are all artifacts of life's long evolutionary heist. The borders we draw between self and other, host and parasite, are blurred at the molecular level. We are mosaics. Every cell in our body carries the memory of ancient intruders. And now, in the Anthropocene, we have become part of the cycle—not just hosts but engineers, not just victims but agents. With tools like CRISPR and synthetic gene drives, we are both observing and editing evolution, echoing the same disruptive creativity that parasites have wielded for billions of years.

If evolution is a story of struggle and innovation, of conflict and convergence, then perhaps parasites were its first artists, working not with canvas or clay, but with code. They taught life how to remix, to borrow, to rewrite. They revealed that survival does not always come from slow change but sometimes comes from a well-timed theft. Now, standing at the edge of a new biological era, we must ask ourselves: What kind of artists will we become?

4

MASTERS OF MANIPULATION

Men are deceived in that they think themselves free; which opinion consists in this, that they are conscious of their own actions, and ignorant of the causes by which they are determined. This, then, is their idea of liberty—that they do not know any cause for their actions.

—Baruch Spinoza, *Ethics*

In college, I was very close to switching my major from biology to philosophy.

Like many students in the early 2000s, I entered college intending to pursue my passions—namely, marine science. My university was Catholic-affiliated, and all students were required to take a minimum of two years of philosophy and theology coursework. I thoroughly enjoyed both. I was drawn to the intellectual rigor, the idea that universal truths might exist (even if some are harder to reach than others), and the continual probing necessary to flesh out the ethical implications of complex issues. But I also loved opening up hard clams under a microscope and watching their tiny microscopic symbionts swimming around inside.

During my sophomore year, I hit a roadblock that every biology major knows all too well: organic chemistry. I came dangerously close to failing the course, and even if I did not, the low grade would have wrecked my GPA and endangered my chances of getting into my dream graduate program. So, on a freezing Monday morning in the middle of February, I made my way to the dean's office at St. John's University's College of Liberal Arts and Sciences, intent on switching my major. As it happened, that particular day, neither the dean nor the assistant dean was in. The secretary suggested I come back the next day to complete the paperwork (this was 2006, when we still carried around actual pink forms to be signed by our advisers). Whether it was laziness or the sheer busyness of that semester, I never followed through. I stuck with my major, and the rest is history. I probably would not be writing this book if Dr. Jeffrey Fagan, the dean at the time, had not decided to postpone or change his schedule that day. It is these small, seemingly inconsequential moments that can reroute the trajectory of an entire life.

Even then, I continued filling my schedule with philosophy courses, with a particular taste for rationalism. One of my favorite philosophers from that school of thought is the seventeenth-century Dutch secularist Baruch Spinoza, who believed that everything, including human thought and behavior, is determined by prior causes. In Spinoza's view, the very idea of agency is an illusion. This contrasts with modern compatibilists like the late Daniel Dennett, who argue that agency can still be meaningful even within deterministic systems.

Spinoza's view of causality resonates disturbingly well with something I would only fully appreciate later in my scientific career: the ability of parasites to hijack the decision-making machinery of their hosts. What if something external—something microscopic—alters the very self that is supposedly doing the choosing? If a parasite changes your brain chemistry, are you really the one making the decision? Free will is central to how humans understand themselves. Almost every religion makes sacred space for the idea of choice, whether framed as the cause of our downfall or the foundation of our spiritual superiority. But what happens when we confront the idea that dark passengers are lurking in nature, organisms that can hijack the very architecture that makes free will possible?

It changes everything.

THE PUPPET MASTERS OF THE ANIMAL KINGDOM

Before diving into specific case studies of parasite manipulation, it is worth understanding the evolutionary groundwork behind these strange and often disturbing behaviors. When a parasite alters its host's behavior, it is not doing so randomly. These manipulations serve a very specific purpose: to bring the parasite closer to completing a stage in its life cycle (filling in another node in the circuit so to speak). In many cases, this means nudging the host into a situation where it is more likely to be eaten by a predator that serves as the parasite's next or final host—a strategy known as trophic transmission. In other cases, the manipulation is about relocation. Take *Ophiocordyceps unilateralis*, the infamous zombie-ant fungus. It infects an ant and compels it to climb to a precise height, bite into a leaf vein, and die. The fungus then erupts from the ant's body and releases spores from this elevated position—a "fungus-friendly" launch site where temperature and humidity are just right for dispersal. In both situations, the parasite is effectively using the host's body as a tool to further its own survival.

When a fungus forces an ant to climb and die, is the ant making a choice? Or has the very mechanism of choice been co-opted? This kind of behavior is sometimes described as an "extended phenotype"—a term coined by Richard Dawkins.[1] In genetics, a phenotype is the physical expression of an organism's genes: eye color, the ability to digest lactose, the shape of a beak. But genes do not stop at the edge of the body. An extended phenotype is any trait, behavioral or structural, that arises from an organism's genes but affects the world beyond its own body. For parasites, behavioral manipulation is encoded in their genomes just as much as physical traits are. Individuals that are better at hijacking their host's behavior are more likely to complete their life cycle, reproduce, and pass on those manipulative genes. Over evolutionary time, this pressure has fine-tuned some parasites into masters of mind control, often with astonishing precision.

This concept is a key feature of parasites that anyone interested in these critters must understand—not just in the lab, but in the classroom. In my junior-level parasitology course, I do not give a final exam. Instead, my students complete what I call the "parasite design challenge." Working in small groups, they are tasked with inventing an entirely new parasite—or, more precisely, a novel host-parasite system that has never been described. On the surface, it sounds fun and deceptively simple. But the moment they try to map out a realistic transmission pathway, or describe the molecular or behavioral tricks their hypothetical parasite might use to manipulate a host, they realize how wildly complex these systems are. Creativity is part of their grade, but their ideas must also be grounded in biological plausibility, which may get messy fast, because as we have seen, parasites are rule breakers! They regularly upend the basic principles students were taught in first-year biology. After teaching this course for eight years, I have noticed a curious pattern: The parasites my students dream up are almost always dramatic, grotesque, and strangely poetic. This is not because my students are sociopaths (although admittedly there could be one or two who might benefit from some psychoanalysis to at least rule out that possibility); it is because they are finally allowed to indulge in the kind of what-if questions that modern formal science education often sidelines due to its annoying obsession with metrics, outcomes, and the like. They get to tinker with the machinery of life—imagining how one organism might alter the behavior, physiology, or even the fate of another. Without fail, there is one real-world parasite that inspires a lot of these projects: the horsehair worms, formally known as the Nematomorpha. This is a group of parasites that does not just control its host; it drives it to suicide.

Nematomorphs, commonly known as horsehair worms because of their appearance, are closely related to nematodes (roundworms). Although they resemble roundworms at a glance, Nematomorpha are only distant cousins of the Nematoda, both belonging to a broader group of molting animals called the Ecdysozoa. But while roundworms include everything from plant parasites to human pathogens, horsehair worms follow a stranger path: They live as larvae inside insects, then emerge—often fatally—in aquatic environments, leaving behind an empty husk of their former host. Nematomorphs are exclusively parasitic in that every species in this animal group must parasitize another animal at some point in its life cycle to complete its development. Imagine you are a cricket, hopping through leaf litter, sensing vibrations, looking for shelter, mates, and food—the usual rhythm of insect life. But somewhere along the way, something microscopic found its way inside you. Maybe you drank from a puddle that carried parasitic larvae. Maybe you ate another insect that was already infected. You would not know—at least not yet.

Days pass, then weeks. Inside your body, something is growing—coiled, silent, and long. A parasite, a horsehair worm, is weaving itself through your body cavity, wrapping around your organs like a living rope. You do not feel it at first; that's part of its genius. But it is not staying forever. One day, the parasite decides it is time: To complete its life cycle, it needs water, and desperately.

That is when the voices start.

You (the cricket) suddenly feel a strange pull—not hunger, not thirst, a compulsion. You begin seeking out light, reflective surfaces, maybe the glint of a puddle or stream. You approach, even if it feels wrong.

Then you leap.

You are underwater. And just before your body shuts down, the worm bursts from you—long, wet, and horrifying. It swims away as your corpse sinks or drifts, its job done. The adult worm will not feed again. It does not even have a proper digestive system. It lives just long enough to find a mate, lay eggs, and scatter a new generation of invisible hitchhikers into the world.

Without a doubt, this is a rough end for the cricket. The worm does not just hijack the insect's body; it rewires its brain. The host almost always dies, either by drowning or from internal trauma as the parasite bursts from its body. Of course, we cannot say the insect "wants" to die. Suicide, in the human sense, implies conscious intent, something a cricket does not possess. What we are witnessing instead is a case of parasitically induced fatal behavior. The host's death directly benefits the parasite, making it a textbook example of extended phenotype manipulation. If you are brave enough, there are plenty of time-lapse videos floating around online—YouTube, TikTok, and the darker corners of nature Twitter—showing infected insects partially submerged in jars of water. The worm, sometimes nearly a meter long, slowly uncoils from the host's abdomen like a wet shoelace from a broken doll. It is hard not to wince (at least on the insect's behalf).

Compared to their better-known cousins, the nematodes, the nematomorphs have long dwelled in scientific obscurity. Their biology was considered too obscure, their life cycle too gruesome, and their hosts too inconvenient. But that is starting to change. A new generation of researchers, armed with molecular tools, is peeling back the curtain, and what they are finding makes the story of the horsehair worm even stranger. For instance, a study by French researchers used proteomics to track protein expression in both host and parasite, before, during, and after the worm induces its host to seek water. They found changes not only in neurotransmitter-related proteins in the cricket's brain, but also in parasite proteins that mimic host signaling molecules, suggesting chemical hijacking of the insect's central nervous system.[2] Even wilder, in a 2023 study, researchers discovered that some horsehair worms may have stolen genes from their hosts![3] Again, scientists compared gene activity in nematomorphs before, during, and after they manipulated the host's behavior. What they found was astonishing: The worm's own genetic machinery kicks into high gear right when it starts controlling its host. And many of the genes that get turned on during this time look suspiciously similar to the genes found in the insects the worms parasitize. This suggests that horsehair worms may have pulled off something thought nearly impossible: horizontal gene transfer (see chapter 3) between complex animals, a process once believed possible only in tiny microbes. Some of those stolen genes might help the worm mimic its host's brain chemistry, making its commands more convincing. In other words, the parasite may be speaking the host's molecular "language," using the host's own code against it.

And the manipulation does not stop with suicide dives into freshwater. In 2024, a study by Japanese researchers published in the journal *Behavioral Ecology* showed that in mantids, which are sexually cannibalistic (female mantids often consume the males during or after mating), horsehair worms seem to have evolved sex-specific behavioral manipulation strategies. Parasitized male mantids show a marked reduction in mating behavior, possibly to avoid being eaten before the worm can complete its life cycle. Females do not show the same suppression—suggesting that the worm is fine with them feeding on mates if it boosts host nutrition.[4]

Taken together, these discoveries reshape what we thought was just a creepy nature video into something far more profound: a complex, evolutionarily honed manipulation system that may involve borrowed genes, neurochemical mimicry, and sex-specific behavior modulation. Horsehair worms are not just macabre curiosities; they are living coiled-up mysteries, and we are only beginning to understand the full extent of their power. No wonder they have become the muses of senior biology majors!

If behavioral manipulation is the endgame, then the first moves were likely made on a smaller battlefield—not in the brain, but inside the cell. Before parasites learned to hijack fear responses or rewire mating and survival instincts, they evolved the ability to control the host's cellular machinery for their own survival. Enter *Trichinella spiralis*, the microscopic nematode worm that has been known since the early nineteenth century, which we briefly discussed in chapter 3. This worm does not just infect its host; it moves in, rewrites the cell's purpose, and builds a home inside the muscle cell. *Trichinella* is the largest known intracellular parasite. It turns your cells into something else—something that serves it.

Sometimes that story begins over a backyard grill.

It was supposed to be a celebratory family reunion. In the summer of 2022, nine relatives from three different states—Minnesota, Arizona, and South Dakota—gathered around a barbecue. One of them had brought something special: black bear meat from a recent hunting trip in Saskatchewan, Canada. The meat had been frozen for more than a month, long enough, they thought, to kill off any lingering parasites. They marinated it, grilled it with vegetables, and served it as bear kebabs, slightly rare.

Within days, strange symptoms started appearing.

One man, a twenty-nine-year-old from Minnesota, was hospitalized twice over seventeen days with unrelenting fever, swelling around the eyes, and intense muscle pain.[5] Doctors could not figure it out. Remember what I said in chapter 1 about the state of parasitology education in U.S. medical schools. It was not until someone asked about his diet that a clue emerged—bear meat. Soon, others from the reunion fell ill too. Three were hospitalized. Six out of the nine guests had been infected—including two people who had not even eaten the meat, just the vegetables cooked alongside it.

Lab tests confirmed the culprit: *Trichinella nativa*, a freeze-resistant species of parasitic worm that had survived forty-five days in a household freezer, and was still found alive in leftover bear meat more than fifteen weeks later. These worms did not just survive cold; they thrived. Once ingested, they did what they have done for millennia: invade the muscles, hijack the cells, and make them home. Under the microscope, the transformation is stunning. A *Trichinella* larva burrows into a muscle fiber and takes over the host cell's machinery, reprogramming it into a "nurse cell"—a living incubator that supplies nutrients, protects the parasite from the immune system, and shelters it for years. No fancy synapses or brain circuits are needed, just cellular dominance. In terms of epidemiology, *Trichinella* species are the culprits behind trichinellosis, the disease that plagued the unfortunate family in our earlier story. While rare in the United States—the CDC reported just thirty-five confirmed cases and six outbreaks from 2016 to 2022—trichinellosis remains a serious zoonotic threat, particularly for those who consume undercooked wild game.[6]

What makes *Trichinella* so fascinating, perhaps even more than the other parasites we have met so far, is its life cycle—or rather, its life cycles. *Trichinella* is unusual in that it maintains two distinct ecological cycles: one domestic and one sylvatic (wild). In the domestic cycle, the parasite circulates mainly among pigs and rodents. Pigs serve as both the intermediate and definitive hosts, acquiring the parasite by scavenging or through contaminated feed. Rats often act as reservoirs, helping to sustain local transmission. Humans are merely accidental hosts—infected when they eat undercooked pork. The sylvatic cycle, by contrast, plays out in the wild, among carnivores and omnivores like bears, wolves, foxes, wild boars, and even walruses. In these ecosystems, *Trichinella* persists through predation and scavenging—trophic transmission in its purest form. Humans typically become infected when they insert themselves into this natural food web by consuming undercooked wild game, just as those campers did.

Public health guidelines often recommend freezing or thoroughly cooking wild meat to kill the parasite. However, *Trichinella* is not a one-size-fits-all organism. It is incredibly genetically diverse, and that diversity translates into adaptability. For instance, *Trichinella nativa*, a species commonly found in arctic and subarctic regions, has evolved to withstand freezing temperatures. This adaptation means that even meat frozen for months may still harbor infectious larvae, making traditional safety practices insufficient in some cases.[7] Luckily all campers survived the infection; the hospitalized patients were successfully treated with the drug Albendazole.

If *Trichinella* represents the parasite's ability to rewrite the rules of cellular biology, turning a muscle cell into a prison and a life-support system all at once, then *Toxoplasma gondii* takes that same mastery to a whole new level. This parasite does not just alter cells; it infiltrates the brain. It reprograms behavior, and it does so with surgical precision, in ways that scientists are only beginning to understand. Whereas *Trichinella* settles in quietly, hidden in the fibers of its host's tissue, *Toxoplasma* marches straight into the central nervous system and begins manipulating its host's most primal instincts—fear, risk, even attraction. We met *Toxoplasma gondii* in chapter 1 as the single-celled parasite famous for flipping the natural fear response of rodents on its head, but we did not explore in depth just how far this manipulation may go. Once the parasite infiltrates a rodent, it begins a quiet, almost surgical campaign, not just in the brain, but in the immune system. One of its first moves is to hijack the host's immune cells—the body's frontline defenders that normally identify and destroy invaders. These cells, which are supposed to help control infections, end up doing the opposite: They carry the parasite deeper into the body, even into the brain. It is as if the body's security guards have been tricked into chauffeuring the interloper straight into headquarters.

For those of you who are somewhat familiar with the human immunodeficiency virus (HIV), this may sound reminiscent of it, as it also targets immune cells. But there is a key difference: HIV destroys these cells to cripple the immune system, whereas *Toxoplasma,* parasite *par excellence,* keeps them alive, altering their behavior just enough to turn them into Trojan horses. The result is not immunosuppression, but redirection—an immune system that is still active but subtly misled. That misdirection gives the parasite the time and access it needs to set up shop in the brain, the decision-making machinery of an animal, where the real manipulation begins. There, it reprograms behavior. Infected mice lose their fear of cats, but more than that, they show signs of increased boldness, greater exploration, and reduced anxiety[8]. They become more likely to enter open spaces, more curious about predator scents, and less hesitant overall. This shift is not a side effect of illness or fatigue; it is a behavioral makeover designed with one goal—getting the mouse eaten.

What happens when the parasite finally reaches its endgame—the feline host? In cats, *Toxoplasma* performs no manipulation. It does not need to. The cat is its final destination, the only host in which it can reproduce sexually and give rise to the next generation of genetically diverse progeny. All the behavioral meddling we see in rodents is not about causing chaos for its own sake; it is about reaching the cat. This is the end, the parasite's goal from the beginning—no more tricks, no more rewiring, just reproduction. The transmission continues when infectious oocysts are shed into the environment through the cat's feces. These oocysts are astonishingly hardy, capable of surviving in soil and water for months, even years. In essence, cats become factories and distributors. While the parasite's manipulation in rodents is finely tuned, its strategy in cats is more infrastructural: Exploit the host's biology to flood the world with infectious particles.

As a cat dad myself, I have often wondered whether some of the odd behaviors I see in my cats might be *Toxoplasma*-related. It has been speculated that infected cats may exhibit subtle behavioral shifts, such as compulsive grooming or altered litter box habits, that could, in theory, enhance environmental contamination.[9] Empirical evidence supporting this is scant, but it is an intriguing possibility. For the most part, though, cats are believed to remain asymptomatic carriers. They are the parasite's final destination, its biological endgame, and usually come through the infection without any noticeable signs. But every rule has its exceptions. In one rare and unsettling case, a six-year-old male neutered domestic cat was brought into the Koret School of Veterinary Medicine with a severe case of clinical toxoplasmosis, exhibiting jaundice, vomiting, and lethargy.[10] Upon veterinary investigation, *Toxoplasma* was found actively infecting multiple organs—including the liver and pancreas. This was not the usual story of a passive host quietly shedding oocyst; this was a body under siege, and tragically, the cat did not survive. What made this case different was the cat's immune status. The severe symptoms were likely the result of immunosuppression, allowing the parasite to replicate unchecked and invade tissue it normally would not. In most healthy cats, the immune system keeps the infection in check, limiting both symptoms and damage. But when that balance is disrupted, even a parasite designed to be a silent partner can become an outright invader. While cases like this are rare, they underscore a broader truth in parasitology: Manipulation, pathology, and outcome often hinge not just on the parasite, but on the condition of the host.

Besides *Toxoplasma*, other similar parasites also employ behavioral nudges to speed up their journey to the next host. Take acanthocephalans, charmingly nicknamed "thorny-headed worms." These parasites infect freshwater shrimp and alter their behavior so drastically that the shrimp begin swimming toward light and spending more time in open water, exactly the sort of risky behavior they would normally avoid. The result? They become easy targets for fish, the parasite's final host. A similar strategy plays out on land.

Certain trematodes infect the eyestalks of terrestrial snails, causing the appendages to pulsate and mimic the appearance of caterpillars. These zombified snails are then more likely to be eaten by birds, completing the parasite's life cycle in spectacularly gruesome fashion. But while these are all clever examples of indirect behavioral rewiring, the insect world offers something even more extreme.

Many horror movies that use parasites as inspiration revolve around a common theme: loss of control. In *Alien* (1979), a facehugger implants an embryo in a human host, which remains alive with no outward signs of infection—until the iconic chest-bursting scene that has been endlessly memed and parodied. In *The Thing* (1982), an alien life-form infects organisms at the cellular level, perfectly mimicking them while sowing paranoia and destroying trust. And in *Slither* (2006), infected individuals fall under the control of a parasitic hive mind, with one host becoming the incubator while others serve as drones. In all of these examples, the parasite is an intruder—an unnatural outsider that infiltrates our world to sow chaos. But then came *The Last of Us*, one of the most scientifically grounded of the genre. Based on the award-winning video game, the story features a mutated *Cordyceps* fungus that infects humans, hijacking their nervous systems and turning them into aggressive, mobile vessels for fungal spore dispersal. The creators consulted real scientists, including behavioral ecologist David Hughes, to make the scenario as plausible—and terrifying—as possible. These hosts are not just infected; they are puppets animated for the parasite's reproductive success. But as chilling as these fictional scenarios are, they still do not match the sheer biological horror of what happens in nature: In the insect world, parasites do not just hijack behavior; they sometimes compel their hosts to protect the next generation of parasites, even after the parasite itself is gone.

Meet the parasitic wasps that transform caterpillars into living, dying bodyguards.The most famous of these wasps by far is *Cotesia glomerata*, only about four millimeters long, roughly the size of a sesame seed. Despite its tiny size, its impact on host behavior is disproportionately large. It begins innocently enough: a cabbage white caterpillar munching leaves under the dappled light of a summer garden. But when *Cotesia* arrives, it is not to feed, it is to implant. The wasp injects dozens of her eggs directly into the caterpillar's body. What follows is a quiet invasion. Days pass. The caterpillar continues its life unaware, while dozens of larvae slowly develop inside it. Then, in a moment both tragic and calculated, the wasp larvae chew their way out. The caterpillar survives the event—barely. What is remarkable is what happens next. Instead of fleeing or dying, the caterpillar remains. For days, it hovers motionless beside the brood. If disturbed, it thrashes violently—a last act of protection. It does not metamorphose. It simply guards, and eventually it dies. This is not a one-off freak event; it is a repeated, evolved pattern of postinfection behavior. The caterpillar's life has been repurposed. It has become a bodyguard for the offspring of its killer. What is even more chilling is that this sacrifice may serve multiple roles. Recent studies suggest that, beyond defending the cocoons, the dying caterpillar may also act as a decoy, drawing the attention of predators away from the vulnerable brood. In other words, it is not just shielding the next generation—it is potentially offering itself up as bait! Most parasitic manipulations are subtle nudges, small behavioral tweaks that help the parasite find its next host. *Cotesia*, however, demonstrates parasitic control on a different level—a complete rewriting of the host's biological destiny.

HOW DO THEY DO IT? THE NEUROBIOLOGY OF CONTROL

Now that we have seen some of the most dramatic examples of parasites manipulating their hosts—from worms that drive insects to watery deaths to wasps that turn caterpillars into dying bodyguards—the natural question is: *How*? How can something so small override behavior, instinct, and even survival? To answer this, we need to peer under the hood of behavior itself. At its core, behavior is not magic; it is biology. It emerges from a complex orchestra of neurons, hormones, immune signals, and gene expression. Our reactions to fear, hunger, sex, light, or danger are all ultimately driven by biochemical circuits: neurotransmitters firing in the brain, hormonal signals circulating through the blood, immune messengers sounding alarms, and genes turning on or off in response. Parasites exploit these systems. They do not reinvent the wheel—they hijack the machinery that's already there. By subtly tweaking neurotransmitters like dopamine, manipulating stress or reproductive hormones, meddling with immune signaling pathways, or even leaving epigenetic signatures on their host's genome, parasites can "puppet" their host from the inside out. Let us take a look at how these manipulations occur, while keeping in mind that, like all scientific mysteries, the nuts and bolts of the mechanisms are still being worked out.

We begin with *Toxoplasma*, arguably the most studied parasite when it comes to behavioral manipulation. One of its boldest tricks involves hijacking the host's dopamine system, the pathway that governs motivation, reward, and, yes, fear. Remarkably, *Toxoplasma* does not just influence the system—it brings its own tools. Biologists have discovered that the parasite carries genes similar to the ones that codes for tyrosine hydroxylase, the enzyme that kickstarts dopamine production.[11] This means that once inside the host's brain, *Toxoplasma* may be cranking up the volume on dopamine—particularly in regions, like the amygdala, that control emotional responses and risk evaluation. The result? A rodent that no longer flinches at the smell of danger, and may even approach it.

But that is only part of the story. In male rodents, *Toxoplasma* also boosts testosterone levels, swelling the testes and enhancing behaviors linked to dominance and sexual signaling. More testosterone means more confidence, more boldness, and, critically, less fear.[12] This hormonal shift also reaches the brain, where it alters the expression of molecules like vasopressin, a neuropeptide that influences social and sexual behavior. In essence, *Toxoplasma* tunes its host's behavior like a radio, turning down fear and turning up the signals that might get the host into trouble, like wandering into open areas or engaging in riskier behaviors. The ultimate prize for the parasite? A better chance of the rodent being eaten by a cat, where *Toxoplasma* can complete the final stage of its life cycle.

But wait, Alice! The rabbit hole goes even deeper . . .

Rather than triggering an all-out immune assault, which could damage both host and parasite, *Toxoplasma* tampers with the host's immune cells, such as dendritic cells and monocytes.[13] These are sentinels of the immune system, but *Toxoplasma* reprograms them to act like chauffeurs, carrying the parasite quietly through the bloodstream to the brain. And here's the eerie part: When scientists look at the brain of an infected rodent, they find the parasite's cysts clustering not at random, but in precisely the regions that control fear and survival instincts. It is like finding surveillance cameras installed only in the most sensitive rooms of a building—*Toxoplasma* knows exactly where to go. The result is something closer to biological engineering than simple infection. This is just one example of how parasites have evolved to manipulate not just the bodies but also the minds of their hosts.

I remember a few years ago, right in the middle of my parasitology lecture, I was deep into explaining how *Toxoplasma gondii* manipulates the behavior of rodents—rewiring their fear response and making them gravitate toward the scent of cats. Then a student raised their hand and, half-jokingly but entirely earnestly, asked: "Could we use that? Like, could we harness whatever *Toxoplasma* is doing and use it to boost performance in athletes?"

There were a few smiles and chuckles in class, but the question stuck with us. It was the kind of query that flirts with science fiction but does not fully step out of scientific plausibility. After all, just a decade earlier, who would have imagined that scientists could transform a bacterial immune system—essentially molecular scissors bacteria use to fight viruses—into the programmable gene-editing tool? A microbial defense mechanism has been repurposed for human design.

What followed was thirty minutes of speculative energy, talking through whether we could ever reprogram a parasite to safely and precisely tweak neural circuits, suppress fear, or enhance risk-taking. In theory, the raw ingredients are all there. *Toxoplasma* already crosses the blood-brain barrier, targets specific regions like the amygdala, and subtly modulates dopamine. It can do what neurologists and pharmaceutical companies dream of doing: influencing behavior from the inside out. But could that be turned into a controllable therapeutic tool? During our class discussion, my students identified four dependencies that would need to be satisfied in order to harness *Toxoplasma* as a behavioral engineering tool. The first is a complete map of how the parasite manipulates host cells at the molecular level—every effector protein, every pathway hijacked. The second is a method of attenuating or "defanging" the parasite so it does not replicate or spread uncontrollably. The third is a delivery system that targets only specific cells or regions in the brain without harming others. Finally, of course, rigorous ethical frameworks would be needed to prevent misuse or unintended consequences. Even with all of this, the unpredictability remains. *Toxoplasma*'s effects vary between host species, between sexes, and even across different strains.[14] It is not a scalpel; it is a biochemical wrench. And unlike CRISPR, it does not come with an easy on/off switch or precise programming instructions. You cannot yet ask *Toxoplasma* to make someone 10 percent braver or 15 percent more risk-tolerant. It is still pretty cool to think about, though.

Now let us switch gears and revisit our parasitic wasps—though it is worth pausing to remind readers that while we often call them parasites, these insects are technically parasitoids. The distinction is key: Whereas parasites tend to keep their host alive, parasitoids ultimately kill them. In the case of *Cotesia glomerata*, the end is not swift. Instead, it unfolds in stages. The caterpillar becomes first an incubator for dozens of wasp larvae and then, astonishingly, a bodyguard for its killers' offspring. It tends to the wasp cocoons, fending off predators until it either starves or is itself attacked.

How does this bizarre guardianship behavior arise?

Interestingly, *Cotesia*'s manipulation of its host does not involve brain invasion. That is what makes this case so fascinating—and unsettling. The wasp injects its eggs alongside a polydnavirus—a symbiotic virus that *Cotesia* has domesticated over evolutionary time.[15] The virus does not infect the caterpillar's brain but instead rewires behavior indirectly, by altering immune and hormonal signaling pathways. Recent studies looking at the caterpillar's internal biology show that its body starts producing different kinds of proteins after the wasp larvae emerge.[16] These changes affect brain chemicals that control alertness and aggression, mess with its stress response, and even disrupt its usual immune defenses. Moreover, the wasp virus suppresses normal hormonal cascades, including those responsible for molting and metamorphosis, effectively freezing the caterpillar in a juvenile, behaviorally pliable state. In other words, the wasp does not need to drill into the brain to exert control. It rewires the system from the periphery, altering the molecules that govern mood, motion, and maturation. The result is an obedient zombie, one that not only tolerates the invader but defends it to the death.

Finally, there is one more parasite that deserves special recognition: the lancet liver fluke, *Dicrocoelium dendriticum*. We touched on it briefly in chapter 1, but its astonishing control over its insect host—the ant—is worth revisiting in greater detail. This tiny flatworm begins life in snail slime, but it saves its most manipulative trick for its second host. Once inside an ant, a single fluke makes a beeline for the brain and encysts itself in a very specific location. What happens next is nothing short of amazing. Every evening, just as the day cools and grazing animals become active, the infected ant climbs to the top of a blade of grass and clamps on tightly with its mandibles. It hangs there, motionless, like a bead of dew, perfectly positioned to be eaten by a passing cow or deer, the fluke's final host. If the ant survives the night uneaten, it climbs back down with the colony, only to repeat the behavior again at dusk. This is behavioral manipulation with surgical precision, timed to the rhythms of both ant and herbivore.

Thanks to recent imaging and molecular studies, we are now uncovering how the fluke pulls this off. The brain-encysting parasite appears to settle near neural circuits responsible for motor control.[17] It could be that infected ants show elevated levels of serotonin—a neurotransmitter associated with repetitive actions and fixated biting. At the same time, levels of octopamine—a neurochemical that drives alertness and escape responses—could also be suppressed, although the exact chemical mechanisms remain unknown. In other words, the parasite does not just park itself in the brain; it chemically rewires the ant's behavioral priorities, dialing up persistence while muting panic. The result is a mindless climber and an immobile sacrifice, perfectly positioned to deliver the parasite into the next stage of its life.

By now, you are probably noticing a trend in the way these manipulators work. Many of them appear to converge on similar strategies, suppressing or increasing certain neurotransmitters in the brain—serotonin to promote fixation or repetitive behavior and dopamine to alter risk-reward systems. Despite their differences in form, habitat, and life cycle, these parasites seem to have independently evolved to hack the same basic circuitry that drives behavior. It is as if nature keeps finding the same pressure points in the brain and pressing them for different ends. Even more fascinating, these manipulations often require only subtle biochemical nudges. The host does not become a mindless zombie overnight. Instead, its behavior is reshaped incrementally, just enough to serve the parasite's goals, without triggering full-blown system failure. Whether it is a cricket leaping into water, a caterpillar standing guard, or an ant climbing to its death, the machinery of control is elegant and disturbingly efficient.

Of course, like any highly publicized area of science, this area of parasitology is not without debate. Not all biologists are convinced that every strange host behavior we see is a finely honed act of manipulation. The question is simple but difficult or, in some cases, impossible to answer: Are the behavioral changes that the parasite is inducing in the host adaptive, shaped by evolution to benefit the parasite? Or are they simply collateral damage, the host's nervous system or immune system malfunctioning under biological stress? Take *Toxoplasma,* for instance. While many studies suggest that it manipulates rodent fear circuits to help it reach its feline host, a few argue that the behavioral changes might just be symptoms of brain inflammation.[18] In other words, is the parasite pulling strings, or is the host just unwell? It is a challenging question, because it requires proving intention without consciousness—demonstrating that these behaviors increase parasite transmission in a repeatable, evolutionarily beneficial way, and are not just byproducts of being infected. This gray zone is where the field of host manipulation currently sits: a mix of clear-cut cases, tantalizing possibilities, and the ever-present need for more data.

What establishes the case for parasite manipulation versus mere pathology? This boils down to a few key scientific criteria that must be met:

1. *Consistent, specific behavioral changes.* The behavior must be specific and predictable, not just a general sign of sickness. For example, infected ants always climb grass at dusk/dawn, infected crickets

leap into water reliably when the worm is ready to emerge, and *Toxoplasma*-infected rodents show selective loss of fear toward cat odors while retaining fear of other predators.

2. *Transmission advantage to the parasite.* The altered behavior must increase the likelihood of the parasite reaching its next host. For example: Rodents losing fear of cats = more likely to be eaten → *Toxoplasma* completes life cycle. Ants clamped to grass = more likely to be eaten by herbivores → *Dicrocoelium* completes life cycle.
3. *Timing and reversibility.* The behavioral change often occurs only when the parasite is ready to move to the next host, and may stop if the parasite is removed or not yet fully developed. This implies strategic timing, not accidental damage.
4. *Mechanistic evidence of manipulation.* There must be sufficient studies that have identified specific molecular, genetic, or neural mechanisms used by the parasite, such as altered neurotransmitter levels, gene expression changes in host neural tissue, cyst formation near motor or decision-making centers, or other similar changes.
5. *Experimental validation.* This is the gold standard: intervention studies in which scientists manipulate either the parasite or the host to test causality. Examples might be blocking a neurotransmitter pathway to prevent the behavioral change or introducing parasite-derived molecules to a host to cause the behavior without the infection.

On the other hand, any of these situations may indicate the opposite—that we are dealing with pathology and not clear manipulation:

1. *Generalized sickness behavior.* This might include symptoms like reduced movement, appetite, or grooming that could just reflect being ill, or anxiety-like or stress behaviors that are not specific to any parasite goal.

2. *No clear transmission benefit.* If the host dies before the parasite transmission occurs, or if the behavior does not actually increase transmission rate, then it is obvious that we are dealing with pathology rather than manipulation.
3. *No mechanistic insight.* If we cannot yet identify how the parasite causes the behavior, or if the changes are highly variable and inconsistent, it may be a case of pathology rather than manipulation. However, the absence of evidence is not evidence of absence. Some mechanisms may simply be more subtle, operating below current detection thresholds. Moreover, on an evolutionary timescale, even erratic behavioral changes might eventually be honed into reliable manipulations if they confer transmission advantages under the right environmental conditions.
4. *Lack of repeatability across studies or hosts.* Finally, a stark red flag would be conflicting results in different labs, or behavioral changes that do not show up in wild populations. This can be the case when novel features emerge as a laboratory artifact rather than a natural occurrence.

In short, if a parasite changes host behavior reliably, strategically, and in a way that benefits its own transmission, and we can trace it to a specific mechanism, it is likely true manipulation. If not, then the behavior may simply be collateral damage rather than adaptive manipulation.

THE HUMAN ANGLE: ARE WE PUPPETS TOO?

It is easy to read about parasitic mind control and feel a satisfying sense of distance. "That's something that happens to rats, or snails, or bugs," we tell ourselves.

But what if the story does not end there?

What if we, too, are being nudged by unseen forces—not with overt puppet strings, but with subtle tugs we barely notice. In the film *The Adjustment Bureau,* a team of otherworldly agents lurks behind the scenes, ensuring that each person follows a prewritten Plan crafted by a mysterious Chairman. If you spilled coffee on your shirt, missed your train, or took a different route home, this was not a coincidence—it was course correction. The smallest detours in your day are engineered to steer you toward your "proper" path. That path might save your life—or simply prevent you from derailing someone else's.

As science peers deeper into the human body and mind, a similar question is emerging, not from fiction, but from biology. Are our decisions entirely our own? Or do they reflect the quiet influence of microbes, parasites, or viruses coexisting with us?

Scientists have begun to ask whether chronic exposure to parasites might influence how we behave, how we organize our communities, and everything else from social conservatism to personality traits.[19] These theories, which sit at the boundary of biology and culture, raise provocative questions about what really drives human values. First proposed by Randy Thornhill and Corey Fincher—trained biologists whose most influential work lies within the field of evolutionary psychology—the parasite-stress theory posits that human cultures evolved different social structures in response to the pressures of local infectious disease environments. In regions historically burdened by parasites, the theory suggests, people developed stronger in-group cohesion, heightened suspicion of outsiders, and behavioral norms aimed at minimizing exposure to unfamiliar pathogens. Closed social systems, religious conservatism, and hygiene taboos may have emerged, in part, as behavioral immune defenses—cultural analogues to the body's immune system.[20]

Empirical studies have lent some support to this. Cross-national comparisons show that societies in high-pathogen regions tend to exhibit greater collectivism, religiosity, and adherence to traditional values.[21] Even within the United States, historical pathogen prevalence has been linked to stronger family ties, more conservative social attitudes, and greater emphasis on conformity.[22] Seen in this light, xenophobia, conformity, and even religiosity may be not just cultural traits but adaptive responses to the microbial threats of one's environment.

Recent research has pushed these ideas further, exploring links between parasite exposure and political ideology. Multiple studies have found that individuals with a history of parasitic infections are more likely to endorse socially conservative beliefs, such as resistance to immigration, traditional gender roles, and punitive justice systems. Interestingly, these effects appear to be driven more by actual past infection than by subjective perceptions of vulnerability to disease.[23] This suggests that lived experience with pathogens—not just fear of them—may shape how people think about order, morality, and governance. These patterns hold across different cultures and remain significant even when controlling for education, income, and age.

One prominent extension of the parasite-stress framework argues that disease burden may help explain why some societies gravitated toward authoritarian political systems, while others embraced more liberal or democratic ideals. In areas where pathogens made life precarious and unpredictable, populations may have favored strong, rule-bound hierarchies that promised stability and social control. By contrast, in healthier regions with fewer infectious threats, people could afford to take more social risks—interacting with outsiders, embracing individual freedoms, and experimenting with new ideas. This theory is supported by modern data showing that countries with a history of high pathogen prevalence tend to score higher on measures of authoritarianism and lower on openness and innovation.[24]

However, not all researchers are convinced. Much of this research emerges from the controversial field of evolutionary psychology, so it is important to consider why this discipline generates both interest and skepticism—context that will help ground the findings that follow. Evolutionary psychology seeks to explain human behavior through the lens of evolution—specifically, by hypothesizing that many psychological traits are adaptations to environmental pressures faced by our ancestors. For example, preferences in mate selection, social behavior, or group dynamics are interpreted as evolved strategies that once enhanced survival or reproductive success. But while this approach is intuitively appealing, especially when applied to broad, cross-cultural patterns, it has drawn sharp criticism from many quarters of the scientific community. A major concern is its heavy reliance on correlational studies. Many claims, such as why we prefer certain facial features or conform to social norms, invoke evolutionary explanations that are inferred rather than directly demonstrated, making it difficult to establish causation.

Furthermore, many hypotheses in evolutionary psychology are difficult to falsify in the Popperian sense.[25] Statements like "men prefer youth because it signals fertility" or "women prefer status because it indicates provisioning ability" are plausible but resistant to empirical disproof. This undermines scientific rigor and has led some critics to dismiss the field as engaging in post hoc storytelling—constructing adaptive narratives for behaviors observed today by imagining selective pressures in Pleistocene environments that we cannot directly access or reconstruct.[26]

Supporters of the field acknowledge these challenges but argue that evolutionary psychology does not rely solely on storytelling. They point to converging evidence from diverse sources—cross-cultural studies, behavioral experiments, twin studies, and even genetics—to bolster the plausibility of their hypotheses. Moreover, they argue that evolutionary psychology is no more speculative than other historical sciences like paleontology or archaeology, which also infer past conditions from present-day traces. But that analogy has limits. While paleontologists can analyze fossilized bones and archaeologists study material artifacts, evolutionary psychologists must infer ancient mental states and behaviors without any direct evidence: Minds do not fossilize, and cognition leaves no physical trace. This epistemological gap makes the field uniquely vulnerable to overreach.

Now, with that context in mind, critics of the parasite-stress theory have raised important methodological concerns—particularly about its heavy reliance on ecological correlations, which compare broad regional traits without demonstrating direct causal mechanisms. Many of these studies, they argue, fail to adequately disentangle pathogen exposure from other confounding environmental factors, such as economic development, education levels, or climate.[27]

Recent reviews have also called for more rigorous theoretical frameworks that distinguish among three distinct levels of influence: (1) the actual pathogen load in a given region; (2) the psychological mechanisms it may activate, such as heightened disgust, sensitivity, or social vigilance; and (3) the long-term cultural or institutional norms that may emerge in response. Without clearly separating these layers, there is a risk of attributing too much explanatory power to parasites alone. In light of these limitations, I would argue that the parasite-stress theory should more appropriately be labeled a hypothesis—provocative and generative, but still speculative and in need of further empirical testing. Even so, the core insight remains intriguing: Parasites may shape not only our bodies but our ideologies and institutions.

On a more individual level, parasites like *Toxoplasma* can influence behavior from within. Much has been made of its effects on humans, especially in pop science and media, where it is often linked to the so-called cat lady syndrome—a supposed condition in which excessive affection for cats and social withdrawal are blamed on latent *Toxoplasma* infection. But this caricature stems from oversimplification and exaggeration. While it is true that *Toxoplasma* can alter rodent behavior, making infected mice less fearful of cats, human effects are far subtler and less deterministic. Research by the biologist Kevin Lafferty of the University of California Santa Barbara, and others, has shown correlations between national *Toxoplasma* prevalence and traits like increased neuroticism, reduced reaction times, higher rates of car accidents, and even entrepreneurial tendencies.[28] Yet these are population-level statistical patterns, not fate inscribed by infection. Recent meta-analyses and studies caution against overinterpreting these findings. They emphasize that while *Toxoplasma* may modulate certain neurological and immune responses, its influence on human behavior is likely small and modulated by many other factors—including environment, genetics, and culture. There is no compelling evidence that infection causes mental illness, personality changes, or obsessive cat ownership. Nevertheless, these studies remind us that even subtle nudges, when multiplied across millions, might shift cultural norms or psychological baselines over time.

PARASITES AND THE PHILOSOPHY OF THE SELF

In his 2022 book *Religion, Disease, and Immunology*, the philosopher Thomas Ellis provocatively argues that religion may be best understood as an evolved cultural system, one shaped by humanity's long entanglement with infectious disease. Practices such as ritual purity, dietary laws, and sacrificial offerings, he suggests, were not merely spiritual symbols; they were adaptive responses to real biological threats. Ellis draws a direct line between bodily immunity and religious belief, framing the latter as a kind of "cultural immune system" designed to detect, manage, and sometimes even expel perceived threats to the community's health cohesion. It is an intriguing proposition, but also an unsettling one. If our deepest metaphysical beliefs and moral structures can be traced back to the threat of parasites and pathogens, what else might be quietly shaping our sense of self?

This brings up a thornier question—one that lies at the intersection of philosophy, neuroscience, and parasitology: *If parasites can subtly or overtly manipulate our thoughts, moods, and behaviors, what does that say about the notion of free will?* In other words, if my aversion to strangers, my vote in an election, or even my general risk tolerance can be nudged by something as small as a microbe, am I truly the one making the decisions? Philosophers have wrestled with the illusion of agency for centuries. Friedrich Nietzsche wrote, "A thought comes when 'it' wills, not when 'I' will." For Nietzsche, the conscious self was not the proud captain of its own ship but a latecomer to processes already set in motion. Parasites give that metaphor biological teeth. They show us that external agents—hidden from perception, yet active in our bloodstreams or brain tissue—might literally commandeer the machinery of choice. Spinoza went even further. He believed that human beings mistake freedom for ignorance. To reiterate the opening quote of this chapter, "Men believe themselves to be free," he wrote, "simply because they are conscious of their actions and unconscious of the causes by which they are determined." What if one of those causes is parasitic? In this light, Spinoza's deterministic universe takes on a new, parasitological dimension. The "self" becomes a battleground of influences, only some of which we can name. Freud, too, hinted at something similar in his model of the unconscious. He argued that much of what drives human behavior lies beneath awareness: primal drives, traumas, and desires, shaped by forces long past. The parasite offers a biological parallel: an invisible puppeteer inside the body, pulling strings the host never feels.

This does not mean that we are mere automatons. But it does suggest that agency is messier, more porous, and more conditional than we would like to admit. The more we learn about behavioral manipulation in nature—whether through *Toxoplasma*'s subtle risk-reward rewiring or the more flamboyant examples like crickets drowning themselves under a hairworm's spell—the more we must confront uncomfortable questions about what it means to be a person. Are we rational agents? Or are we walking ecosystems, our thoughts, feelings, and identities shaped in part by unseen others. The science writer Ed Yong offers a provocative answer in his 2016 bestseller, *I Contain Multitudes*. In it, he shows that the self is not a sovereign entity but a complex assemblage, host to trillions of microbial coauthors that influence everything from digestion to immune response. What we call "me" is, biologically speaking, a committee. Biologists have understood this for decades, but Yong's work reframes the human body for a broader audience—not as a sealed vessel, but as a porous habitat. Just as cities are shaped by their residents, some invited, others not, so too might the mind be shaped by its microbial tenants. If our moods and behaviors can be influenced by changes in our mutualistic microbiome—and, as this chapter explores, by parasites too—then the boundary between self and nonself becomes increasingly porous. We are not lone captains steering our ships; we are ecosystems under negotiation.

PUTTING IT ALL TOGETHER: BETWEEN FREEDOM AND INFECTION

Parasites challenge us to reconsider one of the most sacred tenets of the human experience: our sense of agency. When a worm can make an ant climb to its death, or a microscopic protozoan can dial down a rodent's fear of predators, we are forced to ask: How much of our own behavior is truly ours? The more we learn about the subtle influence of parasites, microbes, and even our own immune systems on cognition and emotion, the more the tidy boundary around the self begins to dissolve. Maybe free will is not a clear-cut power we possess, but a fragile consensus among competing voices—some of them not even human. We flatter ourselves that we are rational agents, masters of our fates. But what if Nietzsche was right? What if a thought is coaxed not by insight, but by something else, like infection?

At the heart of this chapter lies a disturbing but liberating idea: We are not solitary minds; we are walking ecosystems. The forces that shape our choices—our fears, our beliefs, even our politics—may sometimes originate from creatures we cannot see and influences we cannot sense. Perhaps *Invasion of the Body Snatchers* was not so far off—only the real invaders are not aliens or zombies. They are earthly beings that opted for drastically different evolutionary routes than we, and their method is not violent takeover but quiet, chemical suggestion—just subtle enough to let us think it was our own idea all along. When Carl Zimmer wrote *Parasite Rex* more than two decades ago, he helped awaken the public to this bizarre, macabre genius of parasites—their ability to rewrite the rules of biology from within. Two decades later, science has caught up with many of his intuitions—and in some cases, outpaced them. We now know that parasitic manipulation is not just possible but pervasive, and that its effects ripple across everything from animal behavior to cultural norms. And yet, the story is more complex than even Zimmer imagined: subtle, contextual, and shaped by coevolutionary feedback loops we are only beginning to map.

If parasites can hijack the minds of their hosts, one might wonder why evolution has not closed the door. Why have species not found a way to shut them out entirely, to outwit these biochemical saboteurs once and for all? The answer lies not in failure, but in a relentless evolutionary contest. For every trick a parasite evolves to bend its host's behavior, the host develops a countermeasure—a sharper immune system, a behavioral reflex, or a molecular firewall. What emerges is not a one-sided manipulation, but a dynamic arms race, one waged silently over millennia. From genes to ecosystems, parasites have been shaping their hosts just as much as they have been exploiting them. In the next chapter, we step into this hidden battlefield, where parasites do not just deceive, but provoke resistance, resilience, and revolution. This is the story of the arms race that never ends, and the surprising ways it has molded life as we know it.

5

THE HIDDEN ARMS RACE

Now, here, you see, it takes all the running you can do, to keep in the same place. If you want to get somewhere else, you must run at least twice as fast as that!

—Lewis Carroll, *Through the Looking-Glass* (1871)

On the shores of a quiet New Zealand lake, biologists bent over trays of mud and water, sorting through snails no bigger than a fingernail. To a casual observer, the creatures looked identical—tiny brown shells, half-buried in silt, grazing silently on algae. But what the scientists saw was a split population living very different evolutionary lives. Some of the snails reproduced sexually, mixing their genes each generation, while others cloned themselves asexually, producing daughters that were near-perfect genetic copies of their mothers.[1] At first glance, the cloners seemed to hold the advantage: Why waste energy searching for a mate, or producing males, when you can double your reproductive period by going it alone? This rhetorical question usually gets its fair share of chuckles in my undergraduate-level genetics course.

Yet, when researchers followed these populations over time, a pattern emerged. The clones were repeatedly decimated by parasitic flatworms that invaded their bodies and sterilized them from within. Infections swept through the genetically uniform lineages like wildfire, while the sexual snails—though slower to reproduce—persisted. By shuffling their genetic deck with each generation, they stayed one step ahead of the parasites. The cloners, stuck in evolutionary stasis, were easy targets. It was a vivid demonstration of an idea we first encountered in chapter 1: the Red Queen, running fast just to stay in place. There, we met her as a metaphor—an evolutionary monarch reminding us that survival depends on constant adaptation. Here, in the muddy shallows of a New Zealand lake, we see her command made flesh. Parasites change the rules of the game every generation, and hosts are forced to keep up, or perish.

This is not a marginal skirmish in the story of evolution; it is one of its central engines. Every antibody in your blood, every fever that scorches an infection, every shiver of disgust you feel at spoiled food is a product of this endless duel. Parasites evolve new disguises, shedding surface proteins like masks at a masquerade. Hosts counter with ever more elaborate immune systems, inventing genetic diversity as a defense. Parasites retaliate again, finding fresh ways to infiltrate, suppress, or misdirect immunity. And so the cycle continues, an arms race with no ceasefire, shaping the trajectory of life itself.

In this chapter, we leave the metaphor behind and dive into the mechanics of the Red Queen's race. We will see how parasites have sculpted immune systems, from the simplest invertebrates to the most complex vertebrates. We will explore how they may have driven the very invention of sex, a costly but powerful defense against infection. We will trace their fingerprints in the behaviors of animals, from primates grooming lice to chimpanzees avoiding sick troop members. The thesis is simple but radical: Parasites are the hidden adversaries that sharpen life's edge. Without them, our immune systems would be weaker, our behaviors simpler, our ecosystems poorer.

The evolutionary struggle between parasites and hosts has repeatedly forged biological innovation.

REVISITING THE RED QUEEN: FROM METAPHOR TO MECHANISM

Back in chapter 1, we met the Red Queen as a metaphor: Lewis Carroll's monarch running as fast as she could just to stay in place. At the time, she was a playful image for the treadmill of evolution, but here we look at her as something far more concrete. In the endless duel between parasites and their hosts, the Red Queen is not just a symbol of futility; she is the rulebook, dictating the speed at which life must change to avoid falling behind. This is not just about poetry; it is a molecular fact, playing out in immune systems, in genomes, and in the most intimate exchanges between predator and prey at the cellular level.

The idea was first proposed in 1973 by the evolutionary biologist Leigh Van Valen, a direct academic descendent of Theodosius Dobzhansky, one of the fathers of the "modern synthesis" that reconciled Darwinian evolution with Mendelian genetics. Van Valen published a deceptively simple observation: In the fossil record, extinction risk seemed constant, regardless of how long a lineage had already persisted. Dinosaurs, mollusks, trilobites—the odds of disappearing tomorrow looked much the same whether a group had lasted for five million years or fifty. To Van Valen, this suggested that species are not competing against a static environment but against a moving target. Every advance by one lineage ratchets up the pressure on others; to maintain fitness is to run. He called it a "law of extinction," but it was soon reframed as the Red Queen hypothesis, after Carroll's exhausted monarch.

The hypothesis found its champion in William D. Hamilton, a British evolutionary biologist who saw in it a possible solution to one of evolution's deepest puzzles: Why sex? Sexual reproduction, after all, is messy and costly compared to cloning. Hamilton suggested that in a world teeming with parasites, shuffling genes each generation gave hosts a way to stay ahead.[2] By producing genetically diverse offspring, sex created a shifting landscape of resistance alleles, frustrating parasites that had adapted to last year's genotype. Here the Red Queen's logic was not an abstract treadmill but a practical defense strategy: sex as a bulwark against parasites.

Over the next decades, evidence accumulated that host-parasite dynamics do indeed bear the Red Queen's fingerprints. And what once felt like metaphor hardened into mechanism. Nowhere is this clearer than in the phenomenon of antigenic variation—a parasite's ability to endlessly change its molecular disguises. In *Plasmodium falciparum*, the parasite that causes malaria, this is achieved through a family of about sixty var genes. At any one time, only a single var gene is expressed, producing a protein that studs the surface of infected red blood cells. When the immune system begins to recognize that protein, the parasite simply switches to a different var gene, presenting a new target and rendering existing antibodies obsolete. The infection continues, cycling through antigenic masks.[3] Researchers have shown that this switching is not random noise but a highly coordinated system: Parasites follow structured "paths" through antigen space, optimizing their survival while limiting the metabolic costs of constant disguise. Recent genomic studies even suggest that recombination among var genes generates entirely new variants mid-infection, ensuring a fresh repertoire for the next host.[4]

Trypanosomes, the protozoans responsible for sleeping sickness and Chagas disease, take this strategy to its extreme. They cloak themselves in a dense layer of variant surface glycoproteins, switching them stochastically to stay one step ahead of immunity. The host may mount wave after wave of antibody response, but each surge clears only a portion of the population before the next antigenic wave rises. It is like trying to hit a target that constantly changes shape—by the time you recalibrate, the parasite has shifted again. These molecular masquerades embody the Red Queen's logic: perpetual movement, perpetual response.

Hosts, however, are not defenseless. Their countermeasure is genetic diversity, particularly in immune loci like the major histocompatibility complex (MHC). MHC molecules sit on cell surfaces, presenting fragments of parasite proteins to T cells. The more diverse a host's MHC repertoire, the greater the range of parasite antigens it can recognize. Across species, MHC genes are among the most polymorphic (internally different) in the genome, and this diversity is not random. In stickleback fish, females preferentially choose mates with complementary MHC alleles, producing offspring with a balanced, effective immune tool kit.[5] Too few alleles, and the offspring may miss parasite variants; too many, and the system risks misfiring, recognizing self as foreign. Female choice, parasite pressure, and genetic architecture intersect here in a delicate evolutionary equilibrium.

In humans, the story is equally compelling. Our HLA loci, the human version of MHC, are extraordinarily variable, with thousands of alleles maintained in populations worldwide. Studies have shown that certain alleles confer resistance to malaria; others slow the progression of HIV to AIDS. These patterns are fingerprints of parasite-driven selection. Even today, as global health programs combat infectious diseases, the echoes of ancient struggles are written in our genomes. A recent theoretical model suggests that the diversity of HLA alleles is sustained by density-dependent mortality pressures: When an allele becomes too common, parasites adapt to exploit it, reducing its advantage and maintaining variation in the population. This balancing act is the Red Queen rendered in human DNA.[6]

The duel is never one-sided. Just as hosts evolve new recognition capacities, parasites retaliate with faster mutation rates and elaborate switching mechanisms. In *Plasmodium*, recombination among var genes at the chromosome ends generates new antigen profiles, effectively creating a mosaic of disguise. In HIV, high mutation rates and recombination produce antigenic novelty so quickly that immune recognition becomes a moving target. Researchers tracking HIV within single patients see the Red Queen in microcosm: a relentless cycle of recognition, evasion, recognition, evasion, until the immune system is exhausted. What Van Valen saw in fossils, and Hamilton in sex, can now be observed in real time in hospital wards and by sequencing facilities.

Recent studies sharpen this picture further. Work published in *Nature Communications* in 2024 demonstrated that antigenic gene families in *Plasmodium* are maintained not just by mutation but by structured recombination networks that sustain diversity even when the rest of the genome is relatively stable.[7] Another team showed that switching between var genes follows a predictable hierarchy, suggesting that parasites do not merely scramble but plot a course through antigenic space, choosing disguises in a way that maximizes survival odds.[8] In trypanosomes, similar investigations reveal complex epigenetic regulation controlling which variant surface glycoprotein is expressed, hinting at an entire layer of strategy in how parasites pace their disguises. The sophistication is staggering. It is not random sprinting but a choreographed marathon.

Even bacteria are not exempt. Their CRISPR systems, now famous as gene-editing tools, evolved originally as adaptive immune memories against viruses. By storing fragments of viral DNA in their genomes, bacteria can recognize and cut invading phage DNA. Here again is the Red Queen in action: Viruses evolve to evade CRISPR detection, bacteria expand their memory libraries, and the duel continues. From vertebrates with antibodies to microbes with CRISPR, the arms race is universal.

When viewed from above, as in a bird's-eye view, the Red Queen's pace is breathtaking. Parasitic antigenic variation drives host immune diversity; host immune diversity drives parasite innovation in evasion; and the cycle spirals onward. What Van Valen sketched as a general principle is now being confirmed in molecular detail, with genomic tools revealing the real-time choreography of coevolution. It is a dance of reciprocal causation: Each move by one side dictates the countermove of the other. This story is not merely academic. These dynamics shape our daily lives. Malaria's persistence, HIV's slipperiness, even the ongoing challenges of influenza vaccines—all are products of antigenic evolution in parasites and viruses. Our immune systems, honed over millions of years, are monuments to this battle. Without parasites, we would not have the antibody repertoire we rely on, the genetic diversity that buffers us, or even, quite possibly, the system of sexual reproduction.

Unlike the weapons-based arms races that we associate with human civilization, which risks mutual destruction for our species, nature's arms race generates diversity, maintains ecological balance, and fuels innovation. In tropical forests, parasites keep dominant plant species in check, allowing others to persist and maintain biodiversity. In animal populations, parasites prevent monopolies of genotype. They are regulators, not just saboteurs. The Red Queen's race, though exhausting, keeps life varied, dynamic, and astonishingly resilient. From fossils to field experiments, from fish to humans to bacteria, the Red Queen has revealed herself as more than metaphor; she is a mechanism inscribed in DNA, enacted in protein interactions, and made visible in the very structure of ecosystems. And she is still running. Evolution's treadmill never stops, and it is parasites that often set the pace. They are the opponents that force us to adapt and the invisible hands that keep life from stagnating. They are, quite literally, the reason we must keep moving.

IMMUNE SYSTEM INNOVATION: SHAPED BY CONFLICT

The science of immunology has always baffled me, even as a student. The concepts and the mysteries that still exist in this field make it almost divorced from any of the other biological sciences. The stereotypical analogy that we often hear is that the immune system is a fortress: barriers, guards, watchers, scanning for invaders. But in truth it is more like a palace built in enemy territory, evolving under perpetual siege. Every gene, every receptor, every signaling cascade has been hammered and resculpted by the ceaseless pressure of parasites. The immune system is, in many ways, a monument to conflict.

Invertebrates, animals with no bones and no adaptive immunity, lay bare this principle. Their defenses consist of innate systems—phagocytic cells that engulf invaders the way the amorphous amoeba hunts and digests its prey, antimicrobial peptides that punch holes in parasite membranes, reactive oxygen species that attack anything foreign. Yet even with just these tools, many invertebrates live under heavy parasite loads and survive. Mollusks, crustaceans, and insects—each uses variations of the same core strategies. In a snail infected by trematodes, hemocytes swarm into tissues, encapsulating the worm's larvae. A caterpillar attacked by microsporidia may upregulate particular peptides in its gut. The innate system is less precise than adaptive immunity, but in a world of constant parasite pressure, "good enough" is often enough, especially when enhanced by genetic diversity, redundancy, and regulation.

But somewhere in the vertebrate lineage, a radical upgrade occurred: the emergence of adaptive immunity. Suddenly, options exploded: B cells, T cells, recombining receptors, memory—the ability not only to attack but to remember and tailor future attacks. This revolution was not free. It required genome expansions, the development of major histocompatibility complex loci, and mechanisms for distinguishing self from nonself at an even finer scale. But the benefit, in a world crawling with parasites that could mutate or lie hidden, was worth the cost.

Take antibodies—proteins whose variable regions are created by random joining of gene segments in B cells to generate a vast diversity of binding specificities. A single vertebrate can generate billions of distinct antibodies over its lifetime. When a parasite invades, antigens are sampled and presented to T cells, which then help select B cells whose antibody fits best. Those B cells clone, undergo hypermutation to improve binding, and differentiate into memory cells. When the parasite or a related strain returns, the response is faster, stronger, and more precise.

Yet even adaptive systems rely on diversity. Hosts that carry rare alleles—a seldom-seen receptor or a minor variant of MHC—may enjoy temporary protection because parasites have not adapted to it yet. This is frequency-dependent selection in action: Alleles that are common become targets, rare ones become safe harbors, until parasites catch up. Over time, this process preserves genetic variation in immune systems.

The laboratory has provided us with living dramas of this arms race. Richard Lenski's long-term *E. coli* experiments are famous for showing how bacteria evolve over tens of thousands of generations.[9] Less celebrated is the phage-host arms race within those populations. When bacteriophages are introduced, *E. coli* responds via receptor mutations, restriction-modification systems, and other defenses; phages counteradapt to bind new receptors or avoid restriction enzymes. This microcosm echoes the immune battles of vertebrates, albeit in simpler form, reminding us that even in a petri dish, evolutionary conflict is relentless. Perhaps the most literal record of conflict is found in bacteria and archaea: the CRISPR system that we discussed in chapter 3. In this molecular memory bank, short bits of viral (phage) DNA are incorporated into bacterial genomes as spacers. If the same virus tries to attack again, the CRISPR system recognizes the sequence and cuts it. In effect, bacteria write a ledger of past infections. Phages respond by mutating target sequences, deleting or altering protospacers, or deploying anti-CRISPR proteins. It is a genetic arms race writ small—the Red Queen's race in the world of microbes.

In vertebrates, the pressure is multilayered. Parasites evolved antigenic variation; hosts turned to broad receptor repertoires; parasites evolved suppression molecules that blunt immune signaling; hosts evolved countermeasures that detect suppression. In malaria, for instance, as antibodies begin to neutralize one var protein, the parasite switches to another; if hosts improve their MHC presentation or antibody binding, the parasite may counter by downregulating antigen expression or camouflaging itself via binding to host molecules. The dance never ends.

Modern field and genomic work shows that immune genes really do track parasite pressure. In stickleback fish, parasite load is lowest at an intermediate, "just right" number of MHC variants, and experimental exposures show that MHC alleles rise or fall with the specific parasites present.[10] Similar patterns appear in birds, where particular MHC types are tied to avian malaria risk, and along broad geographic gradients, where MHC diversity covaries with parasite exposure.[11] In humans, classic and modern studies show that malaria and viruses such as HIV have left clear signatures on HLA genes, while parasites like *Plasmodium* and trypanosomes evolve fast-changing surface antigens that keep immune systems chasing a moving target.[12]

Thus, the immune system is not a static shield. It is a battlefield ever morphing under pressure. What we call "health" is really just a fleeting truce—a momentary equilibrium between host and invader, vigilance and vulnerability. That balance is delicate, always susceptible to renegotiation by the next parasite variant. Over evolutionary time, this arms race has sculpted not only our defenses but our sense of what it means to be alive: to endure, to adapt, to persist in a world that is always changing its rules.

That tension, between resistance and restraint, lies at the heart of the work honored with the 2025 Nobel Prize in Physiology or Medicine, recently awarded to Mary Brunkow, Fred Ramsdell, and Shimon Sakaguchi.[13] Their discoveries revealed how the immune system polices itself, maintaining tolerance to the body's own tissues while staying ready to fight genuine threats. Sakaguchi first identified a specialized subset of white blood cells called regulatory T cells, or Tregs, which act as the immune system's peacekeepers. Brunkow and Ramsdell, working independently and later collaboratively, traced the genetic and molecular mechanisms that allow these cells to function—notably the FOXP3 gene, whose mutation leads to catastrophic autoimmune collapse.

In the grand evolutionary sense, their work illuminates a truth this chapter has been building toward: The immune system is not just an arsenal; it is a parliament. It negotiates constantly between competing imperatives: attack and tolerance, defense and self-preservation. Parasites have forced it to evolve weapons of exquisite precision, but equally important, they have forced it to evolve brakes. Without those brakes, immunity turns inward, and the host becomes its own victim. In honoring Brunkow, Ramsdell, and Sakaguchi, the Nobel Committee recognized more than a discovery. It recognized a philosophy of balance, the same balance that has guided evolution from its earliest microbial duels to the modern human immune system. Their work reminds us that survival is not about extermination, but regulation: knowing when to fight, when to yield, and when coexistence is the only viable peace. In that sense, the immune system, like evolution itself, endures not through domination, but through restraint.

PARASITISM AND THE EVOLUTION OF SEX

The pond looks calm, but under the surface, war decides who gets to reproduce. Tiny crustaceans called *Daphnia*, water fleas no bigger than a grain of sand, drift through the shallows, feeding on algae and microorganisms. For most of the year, they reproduce without sex, cloning themselves with perfect fidelity. Each daughter is a mirror of her mother, a seamless continuation of a successful design. It is a strategy of confidence: If life is good, why change?

But then, the parasites come. Microsoporidians invade their translucent bodies, hijacking their cells and draining their energy. The infection spreads like wildfire through the clones, each one as vulnerable as the last. And then, almost as if by instinct, *Daphnia* changes its plan. They abandon cloning and begin to produce males—the signal of sexual reproduction. The ponds darken with gametes and embryos, with life scattering its genetic dice in every direction. Most of the offspring will fail, but a few will carry new combinations of genes, new defenses the parasites have never seen before. For *Daphnia*, sex is not pleasure or romance; it is triage. It is an emergency response to invasion.

In college, I remembered a crude joke that my first-year biology instructor once made: "The only thing males contribute is twenty-three chromosomes and an orgasm." It was a lecture hall of 192 students, and there were a few chuckles. Then one female student, one of the university's top lacrosse players (now a successful veterinarian in upstate New York), raised her hand and said, "No professor, actually it's just the twenty-three chromosomes." That caused the entire amphitheater to erupt with laughter. Granted this was in 2005 and such a comment or exchange would probably not go down as well in 2025, but at its core, that was a joke that evolutionary biologists have struggled with for decades: If males are so costly, why does sex exist at all?

At first glance, the answer seems to lie in the past. Sex originated more than a billion years ago, long before complex animals, when early eukaryotes began swapping genetic material. But the real puzzle is not why it began; it is why it persists. In theory, asexual reproduction should take over. Every clone doubles its lineage, but every sexual organism loses half its genetic contribution with each generation. Yet across the tree of life, sex endures. One answer may be parasites. They act as evolutionary enforcers, maintaining sex long after its invention. In clonal lineages, infection spreads like a rumor: One successful parasite genotype can devastate an entire population of identical hosts. Sexual lineages, by contrast, reshuffle their immune genes each generation. The result is unpredictability. It is the one thing parasites cannot easily adapt to. This dynamic was formalized in the maintenance-of-sex hypothesis, first modeled by Hamilton and later refined by researchers like Matthew Hartfield and Peter Keightley, who in 2012 reviewed dozens of coevolutionary simulations showing that parasite-driven selection can maintain sex indefinitely when infection risk and mutation rates are high.[14]

The logic is elegant: Parasites make genetic diversity worth its cost. A sexual population may pay an immediate price in efficiency, but it gains insurance against extinction. A clonal lineage may boom in quiet times, but the next epidemic can erase its entire existence overnight. The evolutionary balance sheet always tilts toward the system that can adapt faster. Recent work has deepened this picture. Over the past decade, many studies have now shown that host-parasite cycles do more than maintain sex; they shape its tempo and geography.[15] Populations under strong parasite pressure evolve higher recombination rates, while isolated or parasite-poor lineages drift toward clonality. Even within a single species, the ratio of sexual to asexual reproduction can map directly onto parasite density. Sex, in this light, behaves like an immune response: It surges during crisis, then retreats when the danger passes.

Parasites do not just sustain sex; they shape the differences between it. Across animals, males and females face infection unequally. Males often burn bright and fast, pouring energy into courtship displays, combat, and growth, leaving less to spend on immune defense. Females, whose success depends on survival and the health of their offspring, tend to invest more steadily in protection. A sweeping analysis of more than a hundred animal species found that, on average, females mount slightly stronger immune responses and suffer fewer infections than males—but the difference was subtle, not absolute.[16] In some species, the roles even reverse. What this tells us is that immunity, like reproduction, is tuned by circumstance—by ecology, hormones, and the invisible pressure of parasites. They do more than infect bodies; they help define what it means to be male or female.

Even humans bear the mark. Men are generally more susceptible to parasitic and viral infections, partly due to testosterone's immunosuppressive effects—a cost of sexual selection that favors boldness, aggression, and metabolic intensity.[17] Women, with higher baseline immunity, pay their own price in autoimmunity: diseases like lupus and rheumatoid arthritis are disproportionately female.[18] Our bodies carry the compromises of an ancient evolutionary negotiation between reproduction and defense. On a genomic scale, parasites may even influence how and where recombination occurs. For example, hotspots of genetic recombination cluster near immune-related genes, particularly the major histocompatibility complex (MHC). This is not coincidence: Reshuffling these regions ensures continual renewal of defense diversity. It also underlies subtler behaviors, like mate choice. In humans and other vertebrates, scent preference tends to favor partners with dissimilar MHC alleles—a subconscious mechanism to diversify offspring immunity. Attraction, it seems, may be one of the Red Queen's tools.

There are, of course, exceptions. Some parasites suppress sex altogether.[19] *Wolbachia*, the bacterial symbiont of insects, can feminize males or induce parthenogenesis, bending host reproduction to its will. Here, the parasite does not just exploit the host—it rewrites the rules of sex entirely. These examples remind us that the evolutionary tug-of-war is not about morals or fairness; it is about control. The interplay between parasites and sex may even extend to the origins of aging. The same recombination that defends against infection also accelerates the turnover of alleles, exposing deleterious mutations to selection. The result is a paradoxical benefit: By forcing continuous renewal, parasites may have helped evolution invent not only sex but mortality—a mechanism to make room for new combinations.

Ultimately, parasites act as both executioners and architects. They prune the slow, reward the variable, and ensure that no lineage grows complacent. Without them, the world might be full of efficient clones—stable, uniform, and doomed the moment the environment changed. Instead, we inhabit a planet of endless recombination, a living kaleidoscope whose colors shift each generation. Sex, then, is not merely about propagation; it is a constant act of defense. Every sperm or egg is a wager against extinction, a ticket in evolution's perpetual lottery. Parasites keep the game running. They make sure the dice never stop rolling.

BEHAVIORAL AND SOCIAL ADAPTATION

Jane Goodall once said that what moved her most about chimpanzees was not their intelligence but their tenderness. She saw it in the way a mother stroked her infant's face, or in the careful patience of one chimp picking through another's fur, pausing to inspect and pluck at some invisible speck. To the outside observer, it might have looked like simple affection—a quiet act of care. But Goodall, ever the patient watcher, understood that these gestures carried layers of meaning. They were moments of hygiene, medicine, and politics all at once.

When Goodall first arrived at Gombe in 1960, few scientists imagined that chimpanzees had anything like human social lives. But her field notes, though criticized for not being "objective" in a quantifiable sense, filled with meticulous detail and the humility of someone learning to see, revealed a world where behavior itself was a kind of immune system. Grooming was not just touch; it was survival. By picking lice and ticks from one another's skin, chimps reduced their parasite loads, soothed anxiety, and reinforced alliances. In her films and photographs, you can almost sense the trance of the act: the intimacy of fingers tracing through fur, the soft grunt of satisfaction, the slow choreography of trust.

Goodall's observations transformed more than primatology; they reshaped how we think about the boundary between biology and emotion. She showed that care and health are entwined—that to touch is to protect, to comfort is to heal. Now, in the wake of her passing, her legacy feels even larger. Every primate groomed in a zoo, every student who studies social bonds and disease transmission, is working in the shadow of her insight—that behavior itself can be medicine.

The Gombe chimps taught us that survival is a matter not only of fangs and fitness but also of touch and time. Across the tropics, primates spend as much as a fifth of their waking hours grooming one another. They pluck parasites, clean wounds, and cement friendships, trading hygiene for trust. Among baboons, those who receive the most grooming recover faster from injuries and live longer. In macaques, low-ranking females barter grooming for protection. Grooming, in evolutionary terms, is diplomacy conducted through fingertips. It keeps parasites in check, but it also holds societies together.[20]

Even the great apes know when hygiene fails. Goodall described how certain chimps avoided individuals visibly infested with lice or suffering from mange, giving them a wide berth during outbreaks. Later studies of mandrills and other primates have confirmed this instinct as a universal one: The smell of infection, subtle as it may be, carries social consequences. Parasites, it turns out, enforce manners. They shape not only who touches whom but who is touched at all. When researchers at the French National Center for Scientific Research (CNRS) studied wild mandrills, they discovered that parasitized individuals, especially those carrying gut worms, gave off distinct odors resulting from chemical shifts in their skin microbiome.[21] Others in the troop avoided them, reducing grooming and social contact until the infection cleared. What looked like ostracism was actually infection control. These animals were practicing their own form of quarantine long before humans created the word.

But avoidance is only one line of defense. Sometimes, survival requires chemistry. In the forests of Central America, white-faced capuchin monkeys were observed capturing millipedes and rubbing them across their fur.[22] The millipedes secreted benzoquinones, potent natural insect repellents. The monkeys became living apothecaries, coating themselves in toxins that deterred mosquitoes and botflies. In the process, they demonstrated something astonishing: an understanding, however instinctive, of dosage and application. Too much millipede secretion, and they risk poisoning; too little, and the parasites prevail.

Across the world, in East Africa, another primate took a more internal approach. Chimpanzees there were filmed stripping bark from *Aspilia* plants, rolling the rough pith across their tongues, and swallowing it whole.[23] The leaves contain thiarubrine A, a compound lethal to nematodes and intestinal protozoa. When healthy, the chimps ignored these plants; when sick, they sought them out. This was not random behavior—it was self-medication. They had learned, perhaps over generations, that bitterness can be a cure. And they are far from alone. Monarch butterflies infected by protozoan parasites lay their eggs on milkweed species rich in cardenolides, compounds that kill the parasite in the next generation.[24] Wooly bear caterpillars, when parasitized, switch from soft green leaves to alkaloid-rich plants that make their bodies toxic to wasp larvae.[25] Even fruit flies, when infected by pathogens, gravitate toward fermenting fruit with higher ethanol content.[26] It can be argued that parasites do not just threaten survival; they teach it.

If primate grooming is the earliest form of medicine, human disgust may be its cognitive descendent. Disgust is not simply an emotion—it is a biological reflex honed by millennia of parasite avoidance. The wrinkled nose, the gag, the shiver at spoiled food or pus or decay, all are ancient algorithms for staying alive. Neuroscientists have traced disgust to the insula, a brain region that also governs taste, suggesting that our revulsion toward contamination evolved directly from our sense of flavor.[27] What once told our ancestors which fruits were ripe now tells us which corpses to bury.

Anthropologist Valerie Curtis called disgust the "behavioral immune system."[28] It acts before conscious thought, and protected us from invisible threats long before germ theory gave them names. The universality of this instinct is striking. Across cultures, humans use cleanliness rituals, taboos, and hygiene practices to fend off contamination. Hindu ablutions, Islamic wudu, Jewish kashrut laws, the Christian association of cleanliness with godliness—all reflect deep evolutionary logic dressed in spiritual language. In chapter 7, "Hitchhikers of History," I delve into a bit more detail on the possible influences of parasites on human myth and religion. Even our burial practices, often framed as moral or religious duties, likely began as responses to disease. The first humans who covered their dead may have been protecting the living from pathogens. Later, that act became ritual—reverent, symbolic, moral—but its origin was microbial. Civilization's earliest gestures of compassion were, in their own way, acts of epidemic control.

Sometimes the line between care and contagion blurs. During the Black Death in medieval Europe, people instinctively avoided the sick. Those who did not—nurses, family members, priests—paid with their lives. We might call it "social distancing" now; they probably called it fate. Yet the same avoidance behavior appears across the animal kingdom. Vampire bats, when infected, reduce grooming and feeding interactions to avoid spreading pathogens to roost mates.[29] Honeybees, paragons of collective living, practice social immunity: When a brood is diseased, workers uncap the infected cells and carry the larvae away, sometimes sealing off entire sections of the hive.[30] In each case, survival depends on recognizing when connection becomes danger. This tension—the need for community versus the risk of contagion—is the ancient tax on social life. The more connected a species becomes, the more efficiently disease spreads. Parasites exploit closeness, and evolution adapts behavior to contain it. For every evolutionary push toward cooperation, there is a parasitic pull toward caution. Our social bonds are stitched together by microbes as much as by minds. In fact, some of our oldest parasite lineages may have emerged precisely because of early human proximity to other species. A compelling case is the intestinal roundworm *Ascaris*, introduced earlier in the book. Humans are typically infected with *Ascaris lumbricoides,* while pigs host *Ascaris suum.* Genetic and paleoparasitological evidence suggests that after pigs were domesticated, *A. suum* may have jumped into humans (or vice versa), adapting to the new host and eventually becoming specialized. Cross-transmission experiments and molecular studies underscore that the boundary between "pig worm" and "human worm" is porous; some human infections even appear of pig origin in modern settings.[31] That historical host-shift may have woven new threads into our relationship with infection, forcing early humans (and pigs) to negotiate intimacy and parasite risk in novel ways.

Modern epidemiology gives this ancient lesson new language. Superspreaders, contact networks, herd immunity—all echo dynamics first tested in troops of monkeys and swarms of bees. I am a city kid, having spent my formative years as a teenager living in New York City, with my favorite destination in the world being Prague.

But cities, those grand experiments in social living, are also perfect incubators for pathogens. From cholera in Victorian London to COVID-19 in New York City in 2020, the same basic truth holds: Cooperation breeds vulnerability. Yet paradoxically, it also breeds resilience. The same empathy that exposes us to others' germs drives us to develop sanitation, medicine, and care. Evolution's arithmetic allows no free lunch, but it rewards innovation born from necessity.

Goodall's chimps were not just grooming partners; they were teachers. They showed us that care can be contagious too. To groom another is to invest in their health; to tend the sick is to risk one's own. These are the paradoxes of social species, and humans have turned them into ethics. The impulse that makes a chimp pick lice from its mate's back is the same one that makes a nurse clean a wound or a parent stay up with a feverish child. Parasites and pathogens played a key role in driving us to invent compassion, because survival demanded it.

In the 2014 Ebola outbreak, survivors—those with antibodies—became caregivers. Immune women organized community response groups, entering quarantined homes to feed and wash the infected.[32] Evolutionarily speaking, it was the Red Queen race playing out in real time: infection, adaptation, immunity, cooperation. But culturally, it was something greater. A species that once learned to avoid the sick had now learned to protect them. Parasites, then, are not simply scourges of life; they are sculptors of behavior. They have written themselves into our gestures, our instincts, our rituals, and our morals. The grooming hand of a chimp, the self-medicating butterfly, the human shudder at decay—all are evolutionary echoes of infection. They remind us that life's great balancing act—between intimacy and infection, between self and society—is not a human invention. It is an inheritance.

Jane Goodall often described her time at Gombe as "an education in humility." She meant that to study other animals was to recognize the continuity of life—to see our reflection in theirs. Now, looking back on her work, it is hard not to feel that humility anew. Every act of care, every boundary we draw between clean and unclean, every moral code about touch and purity owes something to ancient battles of parasites and hosts. When we reach out to hold a loved one's hand, or recoil instinctively from a sneeze, we are reenacting decisions millions of years old. Parasites taught us where to draw the line, and where to cross it. In the smallest acts of intimacy and the grandest gestures of compassion, we are still running the Red Queen's race—together, and against the microscopic world that made us who we are.

PUTTING IT ALL TOGETHER: THE RED QUEEN'S REDEMPTION

The longer one studies life, the clearer it becomes that evolution is not a peaceful march toward perfection but a ceaseless struggle for balance. Parasites are not footnotes in that story; they are the punctuation. Every heartbeat, every mating ritual, every immune response is, in some way, a negotiation with them. They are the reason sex exists, the reason our antibodies never rest, the reason disgust flickers across our faces when we see decay. To be alive is to be under siege, but also to be sharpened by it. The Red Queen's race, then, is a metaphor not for futility but for resilience. She reminds us that survival demands motion, that stasis is death, that life's brilliance depends on its enemies. Parasites are the shadows that sharpen the blade. Without them, the evolutionary edge of life would dull. In an essay titled "The Serviceberry: An Economy of Abundance," Robin Wall Kimmerer offers an appealing, gentler vision of nature, one built on reciprocity and generosity, a world that thrives through giving rather than taking.[33] It is a hopeful view, and a necessary one. But to see nature only as a gift is to miss half its truth. Life's generosity is born from its violence; its beauty is tempered by its brutality. The Red Queen and the Serviceberry tell the same story from opposite ends of the spectrum—one of endless exchange. Perhaps that is where the next story begins—not in battle, nor in benevolence, but in the uneasy truce between them. The Red Queen slows her run and evolution discovers another kind of partnership: symbiosis.

6

SYMBIOSIS

The Secret Evolutionary Shortcut

I am, because we are; and since we are, therefore I am.

—John Mbiti, *African Religions and Philosophy*

One of my fondest memories as a child was visiting the Buccoo Reef in Tobago, in the Caribbean. I remember that summer vividly. My father had planned a weeklong trip for my sister and me, and I was excited as soon as school let out, not only because it was our first real vacation together but because it was my first time traveling by boat with our own bunks. I was fourteen at the time, and I had read and heard so much about coral reefs, especially the famous one practically in our backyard.

On the day we arrived in Tobago, the air shimmered with heat, well over ninety degrees Fahrenheit. After checking into our guesthouse, we changed into our swimsuits and walked to the beach, where a glass-bottom boat waited to take us and a handful of other passengers to Buccoo. As the boat pushed away from shore, my sister and I elbowed for the best position at the glass panels, craning our necks to see every detail. When the boat slowed and the sandy bottom gave way to coral, I felt like I was peering into another world. Beneath us lay what looked like a living forest beneath the sea. Dense aggregations of coral rose like trees, crowded with darting fish, moray eels slipping in and out of crevices, and even the silhouette of a small shark gliding past.

Our guide narrated what we were seeing, pointing out the names of species, but I hardly heard him. My eyes were fixed on the massive branching structures below. It was as if we were looking down on another planet—alien, self-contained, and dazzling in its variety. I had recently learned in school that coral reefs are among the most diverse ecosystems on Earth, rivaling even tropical rainforests. Their secret lay not in the corals alone, but in their intimate partnership with microscopic algae called zooxanthellae. These dinoflagellates, living inside the coral's tissues, gifted the coral with the sugars they made through photosynthesis. In return, the coral provided them with shelter and nutrients. Together, they produced growth rates and structural complexity impossible for either partner alone. These reefs became nurseries, homes, and feeding grounds for a staggering share of the ocean's life.

After the reef, our boat drifted to the Nylon Pool, a shallow sand-bottom lagoon where my sister and I wrestled in the waist-deep water as the sun burned red on the horizon. But even as we played, my mind stayed with the reef. I was fascinated by the idea of corals being "supercharged," at least that was the word I used in my head, by these tiny algae that had traded freedom for life inside another being.

That year, in my biology class, I had learned this was an example of *symbiosis*, a word that comes from the Greek *syn* ("together") and *bios* ("life"). It literally means "living together." The German mycologist Heinrich Anton de Bary coined the term in 1879 in his treatise *Die Erscheinung der Symbiose* (The phenomenon of symbiosis), defining it broadly as "the living together of unlike organisms." Over time, some biologists narrowed the word to mean strictly mutualism, in which both partners benefit, but de Bary's original sense included all forms of long-term biological association, even parasitism. Today, we have returned to his broader usage: Symbiosis encompasses mutualism, commensalism, and parasitism alike.

But what puzzled me, even as a teenager, was how the balance was kept. What stopped the symbiotic algae from multiplying uncontrollably and overwhelming their coral hosts? What prevented the corals from squeezing the algae harder for more nutrients, tilting the relationship into exploitation? It seemed almost too peaceful, a pact between species in which both sides benefited equally. Evolution, I had been taught, was not about fairness, but about advantage. So why did one partner not take more than its share?

As it turns out, this was largely a pseudoproblem. As scientists, we love to categorize nature, because it provides a systematic way of understanding it. Yet in reality, nature is dynamic, and while categories help create a reference point for trying to explain a phenomenon, they should never be treated too literally. Mutualism, commensalism, parasitism—these tidy boxes we learn in school are less like walls and more like road signs. They point us in useful directions, but the real traffic of life often swerves across the lanes.

I did not know it then, staring through the glass bottom of that boat, but coral biologists were already uncovering just how fluid those categories could be. The coral-zooxanthellae partnership I thought of as a perfect mutualism sometimes falters. When water temperatures rise or when pollution stresses the reef, the tiny algae that once "supercharged" the coral can become liabilities. Their photosynthetic machinery starts producing toxic byproducts, damaging the coral's tissues. In response, the coral expels them in a dramatic act we now call "bleaching." What I had learned in class as a neat example of cooperation was in reality a precarious balancing act, one that could slide toward parasitism when the environment tipped.

That realization reshaped how I thought about biology. It was not just corals. Everywhere I looked, once I started to pay attention, relationships in nature were far less stable than I had been taught. Cleaner fish that pick parasites from the mouths of larger fish sometimes cheat, nipping at their hosts' mucus instead. Fungi that help plants absorb nutrients from the soil can, under nutrient-rich conditions, turn into freeloaders, siphoning carbon without offering much in return. Even the microbes in our own guts, which we now celebrate as part of a "healthy microbiome," can flip to harmful opportunists when given the chance. It began to dawn on me that symbiosis is less like a contract and more like a negotiation, constantly rewritten in the face of changing environments. Evolution does not hand out permanent deals. It rewards flexibility, opportunism, and sometimes betrayal.

This is both humbling and strangely reassuring. The world was not locked into rigid binaries of "good" and "bad," "helpful" and "harmful." It was more like a spectrum, a continuum of interactions in which context mattered as much as identity. What looked like harmony from the deck of a boat was, under the surface, an ongoing tug-of-war with no guaranteed winners. And that, in many ways, is the deeper lesson symbioses teach us. They are not static categories but dynamic relationships, shaped as much by shifting conditions as by the organisms themselves. To understand them, we need to let go of the neat boundaries in our textbooks and embrace the messy, conditional reality of life. The next step, then, is to look at symbiosis not as a set of boxes but as a continuum, stretching from mutual benefit through uneasy truces to outright exploitation. It is on this continuum, slippery and shifting, that much of evolution's creativity has unfolded.

SYMBIOSIS AS A CONTINUUM

Symbiosis, as it turns out, is never as tidy as our textbooks suggest. For generations, biology students have been taught to imagine three bins: mutualism, where both partners benefit; commensalism, where one benefits without harming the other; and parasitism, where one thrives at the expense of its host. These categories seem crisp, intuitive, and easy to diagram—and they are. But the real world has little patience for neat boundaries. Out in nature, relationships are slippery, conditional, and context-dependent. Organisms rarely "sign contracts" that lock them forever into one role or another. They negotiate, betray, and shift allegiances, sometimes in ways that blur the line between friend and foe. Modern ecology has increasingly come to recognize this truth: Symbiosis is not a set of bins but a continuum.[1]

Nowhere is this clearer than in corals, those engineers of tropical reefs that dazzled me as a child in Tobago. For decades, scientists portrayed the coral-algae partnership as a paragon of mutualism. Tiny algae, called zooxanthellae, photosynthesize inside the coral's tissues, feeding their hosts with sugars and other metabolites. In return, the corals provide shelter and nutrients. The story had a pleasing symmetry: two very different organisms collaborating to build one of the most diverse ecosystems on Earth. It is poetic and fits the "nature in harmony" narrative peddled by many naturalists. But since the 2000s, new data have chipped away at this idealized image. As mentioned earlier, when water warms or pollution stresses the system, bleaching events occur, in which the algae are expelled from the coral as a result of adverse metabolic byproducts produced in response to those external stressors. At first, this was interpreted as a mutualism gone sour—a tragic collapse when conditions tipped too far. But additional evidence suggests that it might be more complex than we previously thought. It turns out that some strains of these algal "mutualists" may never have been true partners at all.

A 2008 study published in the *Proceedings of the National Academy of Sciences of the USA* described certain genetically distinct groups of the symbiotic algae that appear to act more like parasites than mutualists.[2] These strains still photosynthesize, but instead of providing abundant energy to their coral hosts, they hoard much of it for themselves. In laboratory experiments and field observations, corals associated with these "selfish" symbionts grew more slowly and showed reduced resilience. The algae, in effect, were extracting resources while giving back less in return—a definition of parasitism that biologists would recognize immediately if the partners were not so entwined. This finding was startling not just because it complicated our picture of coral reefs, but because it suggested that even the poster child of mutualism might actually contain parasitic undercurrents hiding in plain sight.

The implications ripple outward. If corals, those icons of symbiosis, can host parasites disguised as partners, what does that say about other relationships we assume to be cooperative? The lesson here is sobering: There is no such thing as an inherently mutualistic partner; there are only organisms responding to costs and benefits as conditions change. Even within the same species of symbiont, different genetic lineages can push the relationship in opposite directions. A coral colony hosting one strain may flourish; another, only a few meters away, may falter under the weight of a more exploitative partner. Symbiosis is therefore less like a static label and more like a sliding scale. At one end lies outright exploitation; at the other, deep cooperation. Most of the time, organisms hover somewhere in between, nudged one way or another by fluctuations in temperature, nutrient levels, or competition.

This perspective has illuminated many other relationships in nature that we once thought were fixed. Take fig trees and their pollinating wasps. For tens of millions of years, these organisms have depended on tiny wasps that crawl into their flowers to lay eggs while inadvertently pollinating them. This looks like a perfect mutualism—each partner needs the other. But some wasps cheat: They lay eggs without delivering pollen. The fig, in turn, can retaliate by aborting the fruit, killing all the wasp larvae inside.[3] It is less a harmonious contract and more an evolutionary cold war, with sanctions and countermeasures baked in. Depending on the strain of the wasp and the condition of the tree, the same partnership can lean toward cooperation or punishment.

Or consider yucca plants and yucca moths. The moths pollinate the flowers while laying eggs in them. Their larvae feed on some of the developing seeds, but enough remain for the plant to reproduce—a balance of give and take. Yet if moths lay too many eggs, consuming too many seeds, the plant can abort those flowers, sacrificing its own potential offspring just to prevent being exploited.[4] What looks like cooperation is, again, an armed truce, with each partner watching the other's moves.

The marine environment offers even stranger stories. In the deep sea, anglerfish carry bioluminescent bacteria in a flashy lure that dangles in front of their mouths. To the fish, this is an extraordinary hunting tool. To the bacteria, it is a cozy home with a steady food supply. But genetic studies have revealed that these bacteria have lost many genes needed to survive independently.[5] They are so dependent on their host that it is debatable whether the relationship is still mutualistic. If conditions changed, the balance could tip into exploitation or collapse. Symbiosis here is not harmony; it is captivity with benefits.

Even human parasites can slide along the spectrum. Tapeworms and many other helminths for instance, are sometimes regarded as a clean +/– interaction when they infect the intestinal tract of humans. Yet in some communities, light infections were tolerated or even believed to offer health benefits, perhaps by calming immune overreactions.[6] The hygiene hypothesis has revived this idea: Parasites may sometimes act as inadvertent regulators, blunting allergies or autoimmune disorders, an indirect benefit of infection (more on the hygiene hypothesis in chapter 8). A villain in one context can become an uneasy ally in another. These examples remind us that symbiosis is not frozen but conditional. The environment matters. Partners matter. Timing matters. An infection that devastates one species may leave another relatively unharmed. A mutualism that thrives in one climate may flip into parasitism under stress. The spectrum is wide, and life is constantly sliding along it.

Stepping back, the continuum model of symbiosis also resonates with human experience. Think of our relationships—not just romantic partnerships but friendships, alliances, even nations in treaties. They are rarely static. What begins as mutual benefit can, over time, tip toward imbalance as one side takes more than it gives. Some relationships stabilize in a workable truce; others collapse under the weight of exploitation. Context, history, and negotiation shape outcomes far more than any intrinsic moral character of the partners. Biology, in this sense, mirrors sociology. The categories we use, such as mutualist, parasite, or commensal, are not destinies but temporary snapshots in an ongoing ecological dance.

If symbiosis is a continuum today, stretching between help and harm, it may always have been so. Perhaps life itself emerged not from perfect cooperation, but from this same messy dynamic. Perhaps parasitism—the cheater, the exploiter—was there at the very beginning, forcing early molecular replicators floating around in a primordial soup to innovate defenses and alliances. To follow that possibility, we turn to an unlikely figure: a mathematician who, four decades ago, proposed that parasitism was not an accident of evolution but its engine.

PARASITES AT THE DAWN OF LIFE: BREMERMANN'S FORGOTTEN VISION

Of the three natural sciences, biology is arguably the broadest. In American universities, biology students are required to take years of chemistry and physics, while the reverse is rarely true. The reason is simple: life is the most complex system we know. To understand it, one must understand not only the chemical components that make it up and the physical laws that govern their behavior, but also how these pieces assemble into higher-order systems—organisms, populations, ecosystems—that defy easy reduction. Biology demands a systems-level approach that transcends chemistry and physics, even as it is built upon them.

This breadth has always meant that many of biology's most important breakthroughs came not from biologists but from outsiders, such as physicists, chemists, mathematicians, and computer scientists, who brought fresh tools and ways of thinking. The discovery of DNA's double helix is a case in point. It was not just the work of biologists but an interdisciplinary race. Chemists like Rosalind Franklin used X-ray crystallography to reveal the molecule's helical structure; Linus Pauling, another chemist, attempted his own triple-helix model. George Gamow, a physicist best known for his work on the Big Bang, was among the first to speculate how nucleotide sequences might encode amino acids, helping set the stage for cracking the genetic code. James Watson and Francis Crick, though trained as biologists, drew heavily on these cross-disciplinary insights. Biology advanced because the problem was too large for one field alone.

The same pattern repeats across the life sciences. Ecology was transformed when mathematicians like Alfred Lotka and Vito Volterra adapted predator-prey equations from physics. Evolutionary biology absorbed ideas from game theory and economics. Today, computational biology leans heavily on algorithms developed in computer science, while synthetic biology relies on engineering principles. Most recently, one of the biggest breakthroughs in modern biology—the development of the programmable gene-editing tool CRISPR-Cas9—also came from an unexpected disciplinary corner. When Jennifer Doudna and Emmanuelle Charpentier won the Nobel Prize in 2020, it was not in medicine or physiology, as many might assume, but in chemistry. That detail matters: It highlights how discoveries that transform our understanding of life often emerge from the chemistry of molecules, not just from traditional biology labs. CRISPR is now used everywhere, from agriculture to medicine, but its roots remind us that biology's deepest revolutions often depend on insights from outside its own borders. Again and again, biology has thrived on intellectual trespass; breakthroughs occur at the borders where disciplines collide. It is in this spirit that we turn to an interesting but provocative idea proposed more than forty years ago: that parasites may hold the key to understanding the very origin of life on Earth.

Most contemporary theories of *abiogenesis*—the emergence of life from nonliving matter—begin with what is often called the "primordial soup." In this scenario, Earth's early atmosphere, rich in gases like methane, carbon dioxide, nitrogen, and ammonia, provided the raw ingredients. Under the right conditions, which included heat from volcanoes, lightning, and ultraviolet radiation, these inorganic molecules could assemble into simple organic compounds such as amino acids and nucleotides, the building blocks of proteins and DNA. This idea was famously tested in the 1950s by Harold Urey and his graduate student Stanley Miller at the University of Chicago.[7] In a now classic experiment, they built a glass apparatus to simulate the early Earth: a closed loop filled with water, methane, ammonia, and hydrogen, continuously sparked with electrical discharges to mimic lightning. After just a week, the water turned pink and then deep red. Inside, they found amino acids—organic matter born from inorganic chemistry. It was a discovery that electrified the scientific world, showing for the first time that life's molecules could be forged under plausible prebiotic conditions.

From there, origin-of-life research blossomed into competing models. Some proposed that RNA, capable of both storing information and catalyzing reactions, was the first self-replicating molecule.

Others emphasized metabolism-first scenarios, in which networks of chemical reactions predated genetic material. Still others argued that life's breakthrough came when lipids (fats) spontaneously assembled into membranes, creating microscopic compartments, so-called protocells, in which molecules could interact, concentrate, and evolve. Despite differences, most of these models converge on a common principle: cooperation. Mutualistic relationships—molecules working together to stabilize, replicate, and adapt—are usually cast as the key to life's first steps.

In 1983, an unusual paper challenged this orthodoxy. Published in the *Journal of Mathematical Biology* by Hans J. Bremermann, a German-born physicist and mathematician who had wandered into evolutionary biology, the article carried the title "Parasites at the Origin of Life."[8] Rather than mutualism, Bremermann suggested that parasitism—the relentless competition between hosts and exploiters—may have been the true spark that drove early molecular systems toward complexity. The paper slipped quietly into obscurity. In the four decades since, it has been cited fewer than fifty times—a footnote in the sprawling origin-of-life literature. Yet its argument deserves a second look. For if Bremermann was even partly correct, then parasites did not arrive late to evolution's story. They were not side characters that emerged only after life had settled in; instead, they may have been there from the very beginning, written into life's first struggles for survival.

Hans J. Bremermann was not, strictly speaking, a biologist. He was born in Bremen, Germany, in 1926 and came of age amid the chaos of World War II. In the aftermath, with laboratories destroyed and resources scarce, he turned to mathematics because—as he later recalled—"all we needed was paper and pencils." That choice set him on a remarkable trajectory. After earning his doctorate in complex analysis at the University of Munster, he worked at the Institute for Advanced Study in Princeton, where he rubbed shoulders with

Robert Oppenheimer and John von Neumann, programmed one of the earliest computers, and began to see that computation could illuminate biology.

Bremermann's career became a tour de force across disciplines: quantum field theory, complexity theory, optimization, genetic algorithms, artificial intelligence, population genetics, and even the emerging field of neural networks. He anticipated what is now known as Bremermann's limit, a theoretical cap on the rate of computation constrained by the laws of physics—ideas that resonate today in the age of quantum computing. He was also among the first to simulate evolution on a computer, decades before genetic algorithms became a staple of AI research. To his students and colleagues at Berkeley, where he taught for more than three decades, Bremermann embodied the conviction that biology could not be reduced to chemistry or physics alone. Life, he argued, was a problem of information processing, of complexity under constraint.

Many of Bremermann's predictions have since found empirical support. His suggestion that pathogens might drive the maintenance of sexual reproduction, because sex "resets" the molecular clock against rapidly evolving pathogens, was echoed independently by evolutionary theorist W. D. Hamilton and later supported by experimental work in clonal versus sexual fish populations. His early efforts to merge fuzzy logic and pattern recognition anticipated modern machine learning's grappling with uncertainty. Even his musings on the energetic limits of computation foreshadowed today's debates about AI scalability.

It was against this backdrop that Bremermann published his 1983 paper. In it, he argued that the earliest steps toward complexity may not have been forged in mutualistic cooperation, as most origin-of-life theories assume, but in parasitic conflict. This was a startling inversion. Rather than seeing parasitism as a later "degenerate" form of life, Bremerman suggested it could have been the engine that forced primitive replicators to evolve defenses, countermeasures, and ultimately, complexity itself.

Let us take a closer look at this theory.

Bremermann's idea begins with a reversal of everything most scientists thought they knew about the origin of life. From the Miller-Urey experiments onward, the dominant story imagined a kind of primordial harmony: Simple molecules gradually learned to cooperate, forming self-replicating strands of RNA, membranes, and proto-metabolic systems that slowly assembled into the first cells. Bremermann asks us to flip that picture on its head. What if life did not begin with cooperation at all? What if the first spark of complexity was exploitation?

The moment any self-replicating molecule appeared, he argued, it would not have lived in peace for long. Mutations were inevitable, and among them would be cheaters—mutant copies that could hijack the machinery of replication to churn out more of themselves without contributing to stability. In other words, parasitism was not a late-stage complication of evolution; it was present from the first moment a molecule dared to copy itself.Bremermann was not talking about complicated math in the abstract—he was trying to show how fragile the very first steps toward life would have been. Imagine a simple strand of RNA, a kind of primitive molecule, floating in a warm pond billions of years ago. It figures out how to copy itself slowly but steadily. That seems like the start of life. But almost immediately, a problem appears. One of its copies comes out slightly different—a mutant. Instead of building all the parts it needs on its own, this mutant takes a shortcut: it "borrows" the machinery and resources that the original RNA worked hard to set up. Because it can skip steps, it copies itself faster and cheaper. Soon there are more of these freeloading mutants than originals. Unless something changes, the original strand that kept the system stable will be drowned out by cheaters, and the whole process could collapse before life really gets going.

This is what he called the "replicator-parasite dilemma": In any open system of replication, freeloaders will inevitably appear and spread. Where others might have seen a fatal flaw, Bremermann saw the engine of progress. Life could not remain a collection of free-floating, defenseless replicators. Parasites forced hosts to innovate. To survive, replicators had to band together, forming alliances and networks, and eventually enclosing themselves within protective boundaries. Cooperation, in this view, did not precede conflict. It was born from it—a desperate solution to the chaos unleashed by parasites.

One of Bremermann's boldest claims was that this pressure explains why life became cellular in the first place. Naked replicators were doomed, vulnerable to every cheater molecule around them. But once enclosed in lipid bubbles, they gained a fighting chance. These proto-cells acted like fortresses, keeping parasites out and protecting the cooperative replicators within. In this sense, membranes, the very hallmark of cellular life, may have been less about efficiency or communication than about defense. Cells, in Bremermann's telling, were walls against freeloaders. From here, he went even further. The parasite-host dynamic, he suggested, is not merely an origin story but a universal law of life. The arms race between cooperators and exploiters never ended; it simply grew more elaborate. From molecular parasites to RNA viruses, from nematodes to tapeworms, conflict has been as central as cooperation in shaping the tree of life. Parasitism is not an afterthought; it is the thread running through all of biology.

What makes this theory even more remarkable is how well it has aged. Four decades later, laboratory experiments with RNA replicators have confirmed exactly what Bremermann predicted: Cheaters arise almost immediately and often dominate unless systems are compartmentalized. Synthetic biology has shown that lipid vesicles (fat bubbles) can indeed suppress parasites, giving enclosed replicators a survival advantage.[9] And in evolutionary theory, the "error catastrophe"—the tendency of defective copies to overrun open systems—is now widely recognized as a core challenge for the origin of life.[10] In fact, many modern researchers agree that the transition to cellular life required precisely this kind of protection from parasites.

Seen in hindsight, what looked like a provocative fringe idea in 1983 now feels prophetic. Bremermann's vision reframes the story of life's beginning not as a tale of molecules slowly learning to cooperate but as a battlefield where cheaters forced innovation. Conflict, not harmony, may have been the spark that drove molecules to become cells and, eventually, to become us. As mentioned earlier, in the four decades since its publication, this article has been cited fewer than fifty times, which is a very modest number compared to the thousands of references lavished on the RNA world hypothesis, metabolism-first models, or endosymbiosis. By the usual metrics of influence, it is a footnote. And yet, in many ways, it is the definition of what scholars sometimes call a "sleeper" paper: an idea so bold and out of step with its time that it slipped quietly past, only to look prophetic decades later.

There are several reasons why Bremermann's theory never gained traction. First was its sheer conceptual boldness. Most models of early evolution leaned heavily on cooperation: molecules working together to copy, build, or stabilize systems. Bremermann flipped the story upside down and argued that the engine of complexity was not harmony, but conflict. Life, he suggested, began in exploitation: The instant a replicator arose, cheaters followed, and only by defending against them did complexity emerge. That inversion may have been too unsettling for many researchers who were building neat, cooperative models of life's dawn.

Second, his timing was unlucky. In the early 1980s, molecular biology was exploding, but the tools to test his idea did not exist. Comparative genomics was decades away. Experimental RNA replicator systems had only just begun to appear, and no one could yet model parasitism at that scale. Bremermann's insight was elegant, but empirically inaccessible. Without a way to test it, most biologists shelved it as speculation and moved on.

Third, his disciplinary position may have worked against him. Bremermann was not a biologist in the conventional sense but a mathematician and physicist. He published in the *Journal of Mathematical Biology*, a venue respected but peripheral to the core community of origin-of-life researchers. His paper fell between cracks: too biological for mathematicians, too mathematical for biologists. Meanwhile, more straightforward grand narratives, like the RNA world and endosymbiotic theories, were attracting attention precisely because they could be tied to laboratory chemistry and compelling experimental pathways. His parasite-first model was harder to grasp, harder to test, and easier to ignore.

And yet, looking back from today, the resonance is unmistakable. Modern synthetic biology has shown that parasitic replicators emerge almost immediately in open systems. Work with lipid vesicles has confirmed that compartments can suppress these cheaters, just as Bremermann predicted. Genomic studies have revealed over and over again that conflict with parasites has been a driver of innovation, from the evolution of CRISPR systems in bacteria to the genetic scars of malaria in human populations.

If that scientific paper did not take off in its own time, it may yet have its moment. Science has a history of rediscovering overlooked insights once the tools catch up. Just ask the Augustinian friar and father of genetics, Gregor Mendel, whose pioneering work on the laws of inheritance languished in obscurity for more than thirty years before being rediscovered. Fortunately for Bremermann, he did not suffer the same fate as poor Gregor, who died in anonymity during his sunset years. Perhaps Bremermann's article was less a dead end than a time capsule, an idea waiting for the twenty-first century to make it testable.

In the end, Bremermann's vision is both unsettling and oddly comforting. It tells us that conflict and dependency are not aberrations in life, but its foundation. If cooperation exists today, it is because parasites forced it into being. As an epitaph for Bremermann, who spent his career moving between disciplines and defying easy categorization, it feels fitting. He saw in parasites what few others dared to: not just destroyers, but the shadow architects of life itself.

GENOMIC REDUCTION AND THE EVOLUTIONARY TRAP

If symbiosis can be an evolutionary shortcut, it is also an evolutionary trap. The same partnerships that allow organisms to thrive together often narrow their future options. Once a symbiont settles into the protected life inside another organism, its genome begins to wither. Genes that are no longer essential are deleted, trimmed, or transferred to the host. Over time, the once independent partner becomes a genetic shadow of its former self, pared down to the bare essentials, unable to live alone. Referred to as "genomic reduction," this is one of evolution's most unforgiving ratchets.[11] Once the machinery of independence is lost, escape is almost impossible.

Nowhere is this more obvious than in our own cells. Mitochondria, the powerhouses of complex life, were once free-living bacteria. More than a billion years ago, they entered into a partnership with an ancestral eukaryote, trading protection for chemical energy. This kind of intimate merger, one cell living inside another, is what biologists call endosymbiosis. Over time, though, most of the bacterial genes were lost or transferred to the host nucleus. Out of thousands their ancestors carried, human mitochondria retain only thirty-seven.[12] Everything else—DNA repair, protein synthesis, metabolic flexibility—was stripped away. Plastids, the green organelles in plants and algae, tell a similar story. Descended from cyanobacteria that once floated free in ancient oceans, they too have been reduced to a sliver of their ancestral genetic repertoire. Today, no mitochondrion or chloroplast can survive independently. They are captives, their genomes whittled down until freedom is no longer possible.

The process is not just confined to organelles. Consider the pea aphid and its bacterial tenant, *Buchnera*. Aphids feed on sugary plant sap, a diet rich in carbohydrates but poor in essential amino acids. On their own, they would starve. But inside their cells live *Buchnera*, a bacterium that supplies the missing nutrients. This alliance has endured for more than a hundred million years. Yet the genome of *Buchnera* is a ruin compared to its free-living relatives. Whole pathways for DNA repair and metabolism are gone; fewer than six hundred genes remain.[13]

The dependence is mutual. The aphid cannot survive without its bacterial partner, and the bacterium cannot survive outside the insect. This is the ratchet in action: one gene lost, then another, until there is no way back. Other insects tell the same tale. Whiteflies carry *Portiera*, mealybugs juggle a consortium of nested symbionts, each covering for the losses of the other. The result is baroque, fragile partnerships stitched together by absence as much as by presence.

Parasites reveal the extremes of this process. Microsporidia, single-celled parasites related to fungi, have some of the smallest eukaryotic genomes known. The microsporidian *Encephalitozoon cuniculi* manages its entire existence with just 2.9 megabases of DNA and fewer than two thousand genes.[14] To put this in perspective, this is equivalent to 2.9 million bases in its DNA, whereas humans possess a genome of 3.2 *billion* bases, more than a thousand times larger. The microsporidian has lost the genes for amino acid synthesis, nucleotide metabolism, even parts of DNA repair. What remains is a stripped-down tool kit dedicated to one purpose: invading a host and reproducing.

Other parasites push even further. Among apicomplexans (the group that includes the infamous agents of malaria), some relatives have reduced their mitochondria to husks. Gregarines, parasites of invertebrates, have independently lost major components of the respiratory chain multiple times. This chain is crucial because it produces the energy currency of life: adenosine triphosphate (ATP). Some species retain only fragments of the Krebs cycle, a vital part of the chain responsible for contributing to ATP product (and the bane of every biochemistry major's existence); others have no capacity for energy metabolism at all, their mitochondria reduced to tiny cellular workshops called *mitosomes*.[15] These mitosomes have lost the machinery to burn fuel, but they still perform one very basic task: assembling little molecular "nuts and bolts" known as iron-sulfur clusters. These clusters are critical cofactors that other enzymes in the cell need to function, so the organelle survives as a kind of parts factory. Imagine a giant power station that once lit up an entire city being stripped down until all it does is manufacture screws and bolts for other machines. That is what evolution has done to the mitochondria of these parasites—turning a powerhouse into a toolbox. In all of these cases, the organelle has become little more than a name—a vestige of ancestry, stripped of its defining functions.

Why does this happen, and why is it so hard to undo (i.e., the trap)? Part of the answer lies in population genetics. Symbionts and intracellular parasites often exist in small populations where chance plays a strong role. Mutations that delete genes can drift to fixation. Meanwhile, if the host supplies a metabolite, there is no selective pressure to maintain the pathway. Once the gene is gone, reacquiring it is almost impossible. Here's another way to look at this: When an organism no longer needs a tool, nature has a way of taking it out of the toolbox. If a parasite can steal a nutrient from its host, the genes for making that nutrient become useless baggage. Over time, random changes chip away at those idle genes until they disappear entirely. And once a gene is gone, it is almost never rebuilt from scratch. This is the ratchet effect: Like a gear that turns only in one direction, genome reduction is effectively irreversible. Horizontal gene transfer occasionally restores a function, but the wholesale reacquisition of lost pathways is extremely rare. The trajectory is one-way: from independence to dependence, from complexity to fragility.

Recent discoveries suggest that reduction can go further than anyone imagined. In 2025, Ryo Harada from Dalhousie University, together with colleagues from Japan, published a stunning study in which they described a single-celled organism (specifically an archaeon) called *Candidatus Sukunaarchaeum mirabile*, whose genome is only 238 kilobases, the smallest known cellular genome to date.[16] This life-form retains just the machinery to copy its DNA and transcribe genes—barely enough to qualify as a cell. Everything else, from metabolism to cell structure, is missing. The researchers likened it to a "cellular ghost," occupying the gray zone between life and virus. This archaeon represents the extreme of the ratchet. Its genome has been pared down to such a degree that independence is inconceivable. Yet it also highlights the creativity of reduction. By outsourcing nearly every function to a host, *Sukunaarchaeum* has carved out a new ecological role. It may depend utterly on its host, but in doing so, it challenges our definition of what it means to be alive.

Seen across the tree of life, these examples suggest that genome reduction is not an exception but a dominant evolutionary pattern. Yuri Wolf and Eugene Koonin have argued that evolution often comprises two phases: short bursts of innovation that increase complexity, followed by much longer periods of reduction and simplification.[17] Endosymbiosis is the perfect arena for this cycle. A new partnership sparks novelty—ATP factories, photosynthesis, nutrient provision—and then the pruning begins, shaving away every redundant gene until only the bare minimum remains. Reduction is a trap, but it is also an engine. It has given us mitochondria, chloroplasts, intimate insect-bacteria alliances, and minimalist parasites that rewrite the rules of biology. It shows that evolution is not simply a march toward complexity; it is also a relentless process of subtraction, in which sometimes the greatest innovations emerge not from adding, but from letting go.

SYMBIOSIS AS AN ENGINE OF INNOVATION

If genome reduction is the trapdoor of symbiosis, the other side of the story is far more celebratory. Partnerships between organisms have not only stripped genomes bare but also unlocked powers no creature could achieve alone. Time and again, symbiosis has proven to be one of evolution's most potent engines of innovation, producing ecosystems, lifestyles, and even new ways of being alive.

Take coral reefs. Earlier, I described the relationship between corals and their algal partners as an uneasy truce that can falter under stress. But there is another side to the story. That same collaboration is also one of the greatest creative forces in the ocean. By teaming up with microscopic algae, corals can grow faster, build larger skeletons, and rise into massive reef structures. These structures do not just shelter corals; they provide the architecture for entire ecosystems. A reef is like a living city. Fish find nurseries among branching corals; clams and sponges filter the water; turtles and sharks patrol the corridors. Together, reefs support roughly a quarter of all marine species, despite covering less than 1 percent of the seafloor. They act as storm barriers for coastlines and as carbon vaults, locking away vast amounts of calcium carbonate. None of this would exist without the quiet cellular partnership between coral polyps and their algal tenants. In this way, reefs are not just examples of symbiosis; they are monuments to what symbiosis can create. Out of a microscopic bargain grew ecosystems of staggering richness, proving that life's most powerful innovations sometimes begin with the smallest collaborations.

The same principle plays out on land in a far more modest form: the lichen. Lichens are everywhere, though most of us pass them by without a second thought. They form crusts of green or orange on tree bark, sprawling gray patches on rocks, speckled patterns on walls. At first glance they seem unremarkable, but lichens are extraordinary pioneers. They thrive where almost nothing else can, thanks to a collaboration between fungi and photosynthetic partners—usually algae, sometimes cyanobacteria. On their own, the fungi would be helpless, unable to make food. The algae or cyanobacteria, while able to photosynthesize, would struggle to withstand freezing, desiccation, or blazing sun on bare stone. Together, they become more than the sum of their parts. The fungus provides structure and protection, holding water like a sponge and shielding the photosynthetic cells from extremes. The partner cells provide energy. The result is not just a partnership but a new kind of organism, one that can colonize bare rock and slowly break it down, seeding the first soils. Every moss, every grass, every tree that eventually grows in such places owes a debt to these humble symbioses. Lichens remind us that innovation is not always dramatic; sometimes it begins quietly, cracking open a foothold in worlds once thought uninhabitable.

If lichens are pioneers of the wilderness, termites show how symbiosis can open an entirely different frontier: the ability to eat what others cannot. Wood, for most animals, is useless. It is made largely of cellulose, one of the toughest biological polymers, and few creatures have the enzymes to break it down. Termites are no exception. On their own, they would starve in the middle of a forest. Their secret lies in the teeming community of microbes in their hindguts—protozoa, bacteria, and archaea that possess the enzymes to dismantle cellulose, ferment it, and release usable energy. Termites have effectively outsourced digestion to their gut symbionts. What looks like a single insect chewing through timber is in fact a living coalition, an alliance of animal and microbe turning the indigestible into opportunity. This partnership has consequences far beyond the termites themselves. By recycling wood, they accelerate nutrient cycles, fertilize soils, and in many ecosystems act as engineers. In parts of Africa, termite mounds alter water flow and soil chemistry, creating fertile oases where grasses flourish and grazing animals gather.[18] Termites illustrate how symbiosis can spark innovation not just in a species, but in whole landscapes. By allowing insects to digest what was once locked away, symbiosis reshaped ecosystems.

Now, let us plunge back into the sea, to a place even more alien than a reef or a termite mound. For centuries, the deep ocean floor was assumed to be barren—a cold, dark wasteland where nothing could survive. Then, in 1977, a submersible diving near the Galapagos Rift stumbled upon something extraordinary. Around hydrothermal vents, where superheated water rich in sulfur gushed from cracks in the seafloor, researchers found dense communities of life: clams, mussels, crabs, and most spectacularly, towering red-plumed worms that reached more than two meters long. These creatures were thriving in perpetual darkness, thousands of meters below the surface, where sunlight never penetrated. It was a discovery that stunned marine science. The secret, once again, was symbiosis. The giant tube worms (*Riftia pachyptila*) lack a mouth, stomach, or digestive system. They cannot feed in any conventional sense. Instead, they house dense populations of chemosynthetic bacteria within a specialized organ called the trophosome. These bacteria use the energy stored in hydrogen sulfide—abundant in vent waters—to fix carbon dioxide into sugars, in the same way plants use sunlight in photosynthesis. The worms, in turn, supply the bacteria with sulfide and oxygen drawn in through their blood-red plumes. The partnership transforms poison into food, sustaining whole communities of vent organisms in the absence of sunlight.[19]

This was more than a biological curiosity; it was a revelation. Life, it turned out, did not need to be tethered to the sun. Energy could be harvested directly from the Earth itself. The discovery of vent symbioses changed how scientists thought about the boundaries of habitability. If ecosystems could flourish miles beneath the ocean in darkness and heat, perhaps similar systems might exist on moons like Europa or Enceladus, where liquid water and chemical gradients are known to exist. Symbiosis here is not just adaptation—it may be the key to life's persistence in extreme environments, and possibly its origin elsewhere in the universe.

Finally, as humans, we too are involved in critical symbioses with other life-forms that are essentially for our survival. The human body is home to trillions of microbes: bacteria, viruses, fungi, and archaea that together form our microbiome. For much of history, we thought of these microbes only as threats, germs to be eliminated. Modern science has rewritten that view. Our microbiota helps us digest foods we could never break down on our own, synthesizing vitamins and producing short-chain fatty acids that nourish our intestinal cells. They also train our immune system, teaching it to distinguish friend from foe. Children raised in overly sterile environments often develop allergies or autoimmune disorders, suggesting that without early microbial exposure, the immune system loses its compass. More provocatively still, evidence hints that our microbes may shape our minds.[20] Some gut bacteria produce neurotransmitters or interact with the vagus nerve, linking gut and brain. While the science is still emerging, correlations have been found between microbial composition and conditions like depression, anxiety, and even autism spectrum disorders.[21] If true, it means that our microbiome does more than digest food—it may help tune our emotions and behavior.

When you look across these stories—the corals building cathedrals of biodiversity, lichens pioneering bare stone, termites digesting wood, deep-sea worms flourishing in darkness, and the human microbiome shaping our health and perhaps even our mood—a pattern emerges. Symbiosis is not just a biological oddity; it is one of the engines of innovation in evolution. It allows life to cross thresholds it could not manage alone, to exploit new diets, to colonize new habitats, to build new ecosystems. These partnerships show us that novelty often emerges from both competition and collaboration, not from solitary struggle but from shared existence. Evolution's creativity lies as much in bargains as in battles. Symbiosis creates pioneers, architects, recyclers, and perhaps even shapers of thought. Symbiosis has fueled some of the greatest leaps in the history of life, yet it has also opened the door for parasites to flourish, infiltrating the machinery that sustains their hosts. To understand how life diversified, we must recognize this double-edged dynamic—that life's greatest engine of creativity also sowed the seeds of its most persistent invaders.

THE DIFFICULTY OF TRANSITIONING

Symbiosis can feel like a golden bargain: two partners come together, share resources, and unlock powers neither could achieve alone. It builds reefs, cracks open bare rock, fuels insects, and even lets giant worms flourish in the dark ocean. But every bargain has its fine print. Like Faust in the old legend, who sold his soul for knowledge and power, organisms that enter into symbiosis often gain dazzling abilities at the cost of independence. What begins as a handshake can harden into handcuffs. The partnership that once looked like liberation can become a prison. Evolutionary biologists now know that this shift—from flexibility to dependency—is far harder than it first appears. Entering a relationship is easy; escaping it, once the ties have tightened, is almost impossible. This is the dilemma of transition, and nowhere is it clearer than in the parasites that have hitched their futures to their hosts.

In the natural world, transitions are often harder than they look. It is one thing for a free-living organism to strike up an association with another species. It is quite another to give up independence altogether and become trapped in a permanent partnership. This evolutionary shift, from free-living to obligate symbiosis, is not inevitable, even when the partnership offers clear benefits. In fact, as scientists Phuong Nguyen and Minus van Baalen have shown, it is surprisingly difficult.[22] Their models suggest that facultative symbiosis—the "best of both worlds," in which an organism can live alone or with a partner—is often the stable endpoint. Obligate dependency emerges only under narrow conditions: when the benefits inside the relationship are extraordinarily high, or when finding and attaching to hosts becomes so costly that investing in independence is no longer worthwhile.

This explains why so many parasites seem stuck in halfway houses of dependencies. Consider *Vibrio cholerae*, the pathogenic bacterium behind cholera outbreaks. It can devastate human populations, yet it still retains the ability to reproduce independently in aquatic environments. It has not surrendered its freedom entirely. By contrast, *Bacillus anthracis*, the cause of anthrax, has lost that option. It cannot proliferate outside a host; its life cycle is locked inside the bodies it infects. The difference underscores Nguyen and van Baalen's point: Giving up sovereignty is not easy, and lineages only take that plunge when the rewards of association outweigh every alternative. I have seen this same tension in my own work on shell-boring polychaete worms of the genus *Polydora*.[23] Their very name is telling: Drawn from the Greek *poly* ("many") and *dora* ("gift"), *Polydora* can mean "the many-gifted one." In Homer's *Iliad*, Polydora is the name of several minor figures, women whose fates are tied to larger, more powerful forces. It is an oddly fitting name for these worms—creatures whose "gifts" are double-edged, both ingenious and destructive, and whose lives are bound up entirely with the bodies of others. Some species, like *Polydora websteri* and *Polydora hoplura*, are obligate symbionts of oysters; they can only reproduce inside the shells of mollusks. Many of their cousins, by contrast, remain free-living in sediments, or slip opportunistically into burrows already made by *Polydora websteri*, as *Polydora cornuta* does. The distinction is striking. One group has crossed the threshold into full dependency, while the other still keeps a foothold in independence. Even within the genus, there are signs of species-specific preferences that hint at early stages of parasitism: *Polydora caeca*, for instance, has been found only in scallops and never in oysters. As a PhD student, I managed to trick *P. hoplura* into laying eggs inside a glass capillary tube, mimicking the burrow it would normally excavate within an oyster shell. The worm could be fooled into reproduction, but it could not escape its fundamental reliance on a burrowed structure. Experiences like this convince me that these worms illustrate the very early steps of obligate parasitism—a Faustian bargain in progress, in which the worms gain security and resources inside the oyster shell, but at the cost of narrowing their evolutionary options.

The same pattern plays out in the fungal partners of plants. Arbuscular mycorrhizal fungi penetrate root cells and exchange nutrients with their hosts. They are so specialized that they cannot reproduce independently in nature.[24] Their relatives, ectomycorrhizal fungi, are a shade less entangled; they form associations around root surfaces and, like my *Polydora* worms, can sometimes be coaxed into reproduction in the lab. Both groups highlight how difficult it is to slide fully down the slope into obligate dependency. Only when a symbiont extracts consistent, high-value rewards from its host does it cross the point of no return.

Once over that threshold, though, the trap is sprung. Genes for independence atrophy. Metabolic pathways are abandoned. Whole tool kits for self-sufficiency vanish. We see this in our old nematode friend, *Trichinella spiralis*, which can only complete its life cycle inside host tissues. Scientists can sometimes rear it through a few molts in vitro, but without a living host it falters, unable to finish the job. Similarly, nematodes like *Haemonchus contortus* or flukes like *Proctoeces* may be coaxed along in artificial conditions, but only up to a point. Once independence is lost, it is almost impossible to reclaim.

There are rare reversals, and they fascinate evolutionary biologists precisely because they are so unusual. Some flatworms, close relatives of parasitic flukes and tapeworms, remain free-living predators and grazers in aquatic environments. Their existence hints that the parasitic path is not the only one available, that freedom can be maintained in certain lineages. Likewise, some fungi that today thrive as decomposers of dead wood may have descended from ancestors that once flirted with parasitism on plants.[25] These exceptions prove the rule. Escaping dependency is like trying to swim against a current—theoretically possible, but in practice exceedingly rare.

The difficulty lies in the mathematics of trade-offs. Nguyen and van Baalen's models show that every adaptation to finding and exploiting a host comes at the expense of investment in independent reproduction. The more energy a symbiont devotes to locating a host, the less it has left to sustain itself alone. At first, this simply shifts the balance toward association. But if the host consistently delivers higher reproductive payoffs, natural selection will slowly erode the symbiont's capacity for life outside. Eventually, it becomes a prisoner of its own specialization. Independence, once surrendered, cannot be easily reacquired because the genes and pathways that supported it are gone. From a parasite's perspective, this can look like success. *Buchnera* bacteria inside aphids, for instance, are so integrated that they can no longer survive on their own. They have shed most of their genome, trusting their insect hosts to provide missing nutrients. In exchange, they supply essential amino acids that aphids cannot obtain from their plant sap diet. Both partners are locked into a codependency that has lasted tens of millions of years. But what looks like harmony is also fragility: neither can walk away.

The parallels to human society are striking. Independence offers resilience; dependency offers efficiency. The tension between the two plays out not just in evolution but in economies, alliances, and technologies. Just as nations debate whether to outsource critical industries, organisms constantly face the question of whether to keep or discard the capacity for independence. As in human affairs, the path of specialization is easier to enter than to reverse. For parasites, this difficulty of transitioning is both a constraint and a creative force. It limits how far some lineages can go, holding them in facultative limbo. It locks others into exquisitely specialized niches, from the mitochondria in our cells to the worms in our guts. Rare reversals remind us that escape is not impossible, but the overwhelming trend is one-way. Evolution, in this corner of life, behaves less like a free market and more like a ratchet: Once the gears click forward into dependency, the teeth prevent a clean rollback.

SYMBIOSIS AND THE DEFINITION OF LIFE

The more we look at symbiosis, the harder it becomes to say where one organism ends and another begins. All of the symbiotic relationships we have covered in this chapter blur the neat boundaries when we speak of "individuals." Biology's traditional categories—organism, parasite, host—start to feel less like absolutes and more like temporary arrangements. That slipperiness raises an ancient and unsettling question: What, exactly, is life?

The theoretical physicist Freeman Dyson once suggested that life had not one but two origins (once again we see a nonbiologist entering our arena, where the big questions lie). The first was metabolic networks of chemical reactions that harnessed energy and built complexity. The second was informational: the emergence of replicators like RNA and DNA that could carry instructions forward in time. Metabolism without replication would fizzle; replication without metabolism would be sterile. Only when the two intertwined did life, as we recognize it, truly begin. Dyson's perspective is useful here because it reminds us that "life" is not one thing but a braid of processes: chemistry, information, energy, and context.

This ambiguity is nowhere clearer than at the margins of biology, where entities like viruses, virophages, and prions challenge our categories. We have already seen, in chapter 3, that viruses are molecular pirates—packets of genetic information that cannot reproduce on their own but come alive inside a host cell. Are they even alive? If life requires autonomy, the answer is no. But if life requires replication and evolution, then viruses certainly qualify, for they mutate, adapt, and diversify with astonishing speed. Virophages complicate matters further. These viruses parasitize other viruses, hijacking their machinery the way viruses hijack cells, giving us a window into how those parasitic replicators that Bremermann hypothesized more than thirty years ago may have operated at the dawn of life. Prions stretch the definition even more. These infectious proteins contain no genetic information, but can propagate by folding and refolding others into their shape. Each of these examples blurs the supposed boundary between living and nonliving, independence and dependence.

From the perspective of symbiosis, perhaps our mistake is in thinking of individual life as the baseline. Increasingly, ecologists argue the opposite: that symbiosis may be the default condition of life and "individualism" the exception.[26] Virtually all animals exist with microbes, and more than 80 percent of land plants form associations with fungi. Even our cells, with their mitochondria and plastids, are the products of ancient mergers. The neat individual of a biology textbook may be an illusion, a convenient simplification of a much messier reality.

Viruses, too, can be seen through this lens. Instead of asking whether they are alive on their own, we might say that they represent the most stripped-down form of symbiosis: genetic information that only comes into being inside another organism. In this light, viruses are not anomalies but reminders that life has always been relational. Even the smallest self is never really alone. Thinking this way shifts the question from "What is life?" to "Where does life end?" Is an aphid still an aphid without its *Buchnera* symbionts? Is a coral still a coral without zooxanthellae? Are we still fully human without our microbiome? The boundaries are less like walls and more like membranes—porous, conditional, negotiated.

Dyson's dual origins—the metabolic and the informational—map neatly onto this blurred reality. Some partnerships specialize in energy, others in information, still others in scaffolding environments for evolution. Taken together, they suggest that life is not a property locked inside individuals but a process that emerges between them. To define life, then, is not to draw a hard line between "living" and "nonliving," but to recognize a spectrum of relationships, each sustaining the other.

SYMBIOSES IN HUMAN TERMS

The language of symbiosis has always seeped into human affairs. We call exploitative corporations "parasites." We describe marriages as "mutualisms." We speak of political alliances as "symbiotic." These metaphors come easily because the biological realities are so unsettling: organisms that lose their boundaries, merge with others, and surrender part of themselves in exchange for something they cannot provide alone. Faced with such strangeness, we reach for the nearest human analogy.

Writers and artists have long mined this territory. Vladimir Nabokov, who devoted his life to butterflies, saw in their transformations metaphors for hidden lives and metamorphoses of the self.

David Cronenberg, master of body horror, turned symbiosis into nightmare: His films *Shivers* and *The Fly* recast merging with another organism as grotesque violation. Superhero comics offer a different angle: Marvel's *Venom* symbiote is at once parasite and savior, draining its host while endowing him with extraordinary powers. And James Cameron's *Avatar* pushes the metaphor toward reverence: The Na'vi's bond with their animals and their planetwide neural network portray symbiosis as a kind of ecological communion. Whether it horrifies or uplifts, symbiosis is never neutral in our imagination.

Part of the seduction of metaphor is that it makes the alien familiar. Political parasites, corporate hosts, marital contracts—these images give shape to dynamics we otherwise struggle to grasp. Evolution has no intention, no morality, but when one partner exploits another, it is hard not to describe it as cheating. When two species flourish together, it feels natural to call it a partnership. Metaphors offer comfort; they reassure us that these bizarre entanglements map onto something we know. When I teach about symbiosis, I too utilize such metaphors, which makes it easier for students to grasp specific concepts. But here lies the danger. Metaphors can smuggle in human assumptions that biology does not share. A parasite is not malicious. A coral does not "agree" to house algae. A fig tree does not "punish" cheating wasps out of spite. These are biochemical negotiations, not moral dramas. The language of contracts and alliances risks obscuring just how impersonal, and yet how astonishing, evolution is.

Still, the fact that we lean so heavily on metaphors says something profound. Symbiosis destabilizes our most basic categories: self and other, autonomy and dependence, inside and outside. It forces us to confront the idea that individuality itself might be an illusion. No wonder we fall back on stories—whether Cronenberg's horror or Cameron's ecological fable—to process what it means to live not alone, but entangled.

In the end, metaphors may be less about explaining symbiosis than about revealing ourselves. We cast parasites as villains or allies because we cannot help but see our own anxieties in them: fears of invasion, or betrayal, of losing control, but also dreams of transformation and connection. Parasites and symbionts have always been puppeteers in biology. In our metaphors, they become puppeteers of our imagination as well.

PUTTING IT ALL TOGETHER: HOW TO BECOME A PARASITE

In the end, I keep returning to that glass-bottom boat over the Buccoo Reef—the way the water sharpened into focus and a whole city of life appeared where I'd expected only sand. Back then, I thought the secret was simple: Corals plus algae equals a thriving reef. What this chapter has hopefully made clear is that there is always a second story running under the first. The reef was not only a monument to cooperation; it was a high wire strung between help and harm, a balancing act performed moment by moment by partners whose interests overlap just enough to keep the show going.

If symbiosis can unlock new evolutionary possibilities, it also creates instability. Too much self-interest pushes a partnership toward parasitism; too much generosity invites exploitation. This is why so many partnerships we have met feel provisional—lichens cracking open bare rock, termites outsourcing digestion, tube worms turning poison into food, our own bodies hosting microbiomes that can steady our immune system or, under stress, turn unruly. Each is a living negotiation. The terms change with temperature, with scarcity, with the arrival of a new strain or an opportunist. Categories help us talk about what we see, but the life beneath our labels never stops moving.

Seen from that angle, "How to Become a Parasite" is less an instruction manual than an evolutionary pattern—one that repeats so often it deserves to be named. First comes *opportunity*: a leak in the wall, a resource left on the table, a host whose chemistry can be tapped. Then comes *attachment*: some way to get close and stay there, whether as an endosymbiont tucked inside a cell, a bacterium nested in an aphid's organ, or a worm sealed in an oyster's shell. Next is *specialization*: Genes for independence grow idle when a partner supplies what they once provided; selection trims the fat, and random loss finishes the job. Finally comes *lock-in*: With pathways deleted and options gone, the relationship hardens from choice into fate. The handshake becomes handcuffs. From the outside, it can look like success—an exquisitely tuned life-form thriving in its niche. From the genome's point of view, it is a one-way street.

Yet to call parasitism merely a fall from grace would miss the larger truth that this chapter is advocating for. The same road that ends in captivity is the road that gave us mitochondria and chloroplasts, the road that built reefs and soils and vent cities in the deep. Reduction is a trap and an engine at once: Subtract enough and you narrow your future, but in the narrowing you sometimes discover entirely new ways to live. That is why Bremermann's old provocation—parasites at the dawn of life—rings true to me. Conflict did not cancel cooperation; it sculpted it. Cheaters forced walls into being; walls made cells; cells made the deals that built the natural world.

The difficulty of transition, our Faustian fine print, helps explain why the tightrope is crowded in the middle. Many lineages linger in flexible arrangements, able to live alone or together. Few step cleanly into lifelong dependence, and fewer still make it back. The shell-boring polychaete worms that undergrads in my lab study tells that story in miniature: cousins that keep one foot in the sediment and cousins that have crossed the threshold, laying eggs only in the safety of a shell, even when the "shell" is a glass tube in the laboratory. What begins as a preference can gradually become specialization, and specialization can eventually become evolutionary constraint.

If this chapter has a single lesson, it is that parasitism and mutualisms are not separate kingdoms. They are points along a spectrum, and organisms slide between them as conditions shift. The reef that dazzled me as a child is built on generosity and policed by sanctions; the lichen's quiet industry rests on a marriage that would fail without shelter; the termite's bounty depends on passengers in its gut; the tube worm's splendor is a pact with bacteria that turn volcanic poison into food. Even our own selves—the neat "individuals" of textbook diagrams—are composites. We are endosymbiotic chimeras negotiating daily with microbes that help us digest, teach our immune system, and perhaps nudge our moods. Ask where life begins and ends, and you discover that the boundary is porous, like a tide line on a beach.

So, how to become a parasite? You don't set out to. You find a partner because together you can do more than you can do alone. You settle in for the long nights because the inside is safer than the outside. You let unused tools rust because maintenance is costly. One day you reach for a handle you used to turn and discover it is gone. What remains is a narrower life that, paradoxically, may be better at being exactly what it is. Evolution is full of such bargains. Some end in empires; some end in cul-de-sacs. Most end somewhere in between, on the tightrope, where creativity and risk are the same step.

When I picture the Buccoo Reef now, I still see the forest of coral and the darting fish. But I also see contracts, invisible and everywhere—the algae exchanging sugars for shelter, the microbes striking their silent deals, the constant testing at the edge where collaboration blurs into exploitation. If there is an ethic that emerges from this, it is not the moral of harmony or the cynicism of theft. It is the discipline of attention: to relationships, to context, to the ways life makes and unmakes itself together. It is enough to say here that the reef that raised my questions also holds the answer. I am, because we are. And because we are—sometimes allies, sometimes adversaries, always entangled—life keeps finding new ways to go on.

7

HITCHHIKERS OF HISTORY

The world is full of obvious things which nobody by any chance ever observes.

—Arthur Conan Doyle, *The Hound of the Baskervilles*

PARASITES AS SCULPTURERS OF THEIR HOSTS' GENOMES

When we tell the story of humanity, we often spotlight our migrations, inventions, and empires. But trailing alongside us on every step of that journey were far smaller companions—parasites and pathogens. These invisible hitchhikers have quietly shaped our evolution, our movements, and even our cultures. From worms in prehistoric intestines to pandemics that redrew the map of civilizations, parasites have been among the most persistent and overlooked forces in human history. Their influence is inscribed in our DNA. Across deep evolutionary time, recurrent infections acted as a crucible of natural selection, favoring genetic traits that conferred even the slightest edge in survival. Our genomes still bear the scars of these ancient arms races. Thanks to breakthroughs in genetic forensics and biotechnology, we can now trace those scars with remarkable precision.

Modern genomics allows researchers to scan the human genome for signatures of adaptation—regions where natural selection has rapidly shaped our immune responses, blood chemistry, and even metabolism. Perhaps most astonishing is the rise of ancient DNA analysis, which lets us peer back into the genetic past. Ancient DNA refers to genetic material extracted from long-dead remains—bones, teeth, mummies, and even fossilized feces. For decades, such traces were thought to be too degraded to recover. But today, high-throughput sequencing technologies have changed the game, allowing scientists to reconstruct fragments of human and pathogen genomes from individuals who lived thousands of years ago. These fragments are like molecular fossils—clues to the biological struggles that shaped us. By comparing ancient genomes to those of living people, we can map when and where key adaptations to parasitic threats first emerged. The results are striking. Across prehistoric graves and archaeological digs, scientists have identified DNA from parasites like *Plasmodium* and *Trypanosoma*, causative agents of malaria and African sleeping sickness, along with other infectious diseases that left detectable traces in ancient remains. In some cases, parasite eggs or lesions on bones provide physical evidence of infection by *Ascaris*—those large worms that made my students squeal in horror and delight. In others, it is the host's DNA that tells the story: immune system genes that evolved in response to relentless infection pressures. Many of the fastest-evolving genes in the human genome are related to immunity—a genetic echo of how vital parasite resistance was to survival.

Consider malaria, caused by the microscopic single-celled protozoan *Plasmodium*. Despite their tiny size, these organisms have stalled armies, collapsed economies, and stunted the growth of entire nation states. Even today, it is a scourge, with more than half a million people dead from malaria in 2020 alone.[1] But perhaps

Plasmodium's most profound legacy lies not in the headlines of history, but in the human genome itself. In regions of sub-Saharan Africa where *Plasmodium falciparum*—the most lethal of the *Plasmodium* species—has plagued human populations for millennia, causing fevers and hallucinations, natural selection carved powerful genetic defenses into our DNA. The most famous of these is the sickle cell mutation: a single-letter change in the gene that encodes hemoglobin, the protein responsible for carrying oxygen in red blood cells. When inherited from one parent, this mutation reshapes red blood cells into a slightly sickled form—enough to make it harder for the parasite to invade and reproduce, offering a life-saving edge. But when inherited from both parents, it causes sickle cell disease, a painful and often debilitating condition in which deformed cells clog blood vessels and starve tissues of oxygen.

This brutal trade-off is a textbook case of balancing selection, in which the evolutionary benefit of disease resistance keeps a harmful trait alive in the population. But what happens when the selective pressure disappears? Today, descendants of West and Central Africans—many of whom were forcibly brought to North America through the transatlantic slave trade—still carry the genetic legacy of this adaptation. In the malaria-free environments of the United States and other parts of the world, the protective benefit is gone, and what remains is a dangerous burden. Evolutionary biologists refer to this as a maladaptive trait: a once-advantageous mutation that becomes harmful when the environment changes. This underscores how ancient evolutionary battles can continue to shape modern patterns of disease and health inequality, long after the original threat has vanished.

Sickle cell was only the beginning. In West Africa, an even subtler genetic shift spread quietly: the Duffy antigen mutation. By knocking out a single receptor on red blood cells, entire populations closed the door on *Plasmodium vivax*, another malaria parasite. To this day, *P. vivax* is almost absent from sub-Saharan Africa. It thrives elsewhere—in India, in Southeast Asia, in the Americas—but West Africans carry an invisible shield in their bloodstreams, a shield forged by thousands of years of relentless exposure.

Recent advances in ancient DNA research have given us a clearer window into how these battles unfolded. A 2024 study published in *Nature* successfully sequenced *Plasmodium falciparum* DNA from human remains dating back more than five thousand years in Europe, revealing that malaria was far more geographically widespread in the past than previously believed.[2] The ancient parasite genomes also showed long-term evolutionary stability, suggesting that the same selective pressures acting in sub-Saharan Africa were shaping human populations elsewhere. Even in Europe, signs of malaria's evolutionary imprint—such as the rise of protective mutations like G6PD deficiency, a genetic trait that alters red blood cells in a way that offers some protection against malaria but can also make people vulnerable to certain types of anemia—suggest a parallel story of adaptation and cost. These findings not only confirm that malaria drove significant genetic change across continents but also demonstrate the power of ancient DNA to track evolutionary responses to disease across time and space.

Malaria is not the only parasite to leave its mark on our DNA. Thousands of miles away, in the lowland rainforests of the Amazon, another microscopic invader has been waging a quiet, deadly war against humans for millennia: *Trypanosoma cruzi*, the protozoan parasite that causes Chagas disease. Transmitted by blood-feeding insects known as "kissing bugs," *T. cruzi* infects an estimated six to seven million people globally, with millions of cases in the Americas alone. About 30 percent of those infected go on to develop life-threatening cardiac or gastrointestinal complications, often decades after the initial infection.[3] For years, scientists assumed that Chagas was a relatively recent threat, intensifying only in the post-Columbian era. But a groundbreaking 2023 study published in *Science Advances* has radically shifted that view—and opened an astonishing new window into how parasites have shaped human evolution in the Americas.[4] Researchers sequenced the genomes of 118 individuals from nineteen present-day indigenous Amazon populations. What they discovered was a genetic smoking gun: a cluster of immune system genes in ancient people showing signs of strong, recent natural selection. These genes are involved in inflammatory responses and cardiac function—precisely the biological systems affected by chronic Chagas infections.

This is the first documented case of parasite-driven genetic adaptation in the Americas. And it is exciting for several reasons. First, it rewrites what we thought we knew about the disease landscape of the pre-Columbian New World. Far from being disease-free, indigenous populations were battling powerful pathogens long before European contact. Second, it shows that ancient DNA can reveal the footprint of parasites that leave no visible trace in bones or artifacts. Where archaeology is silent, genomics speaks volumes. Most compelling of all, the protective variants shaped by those ancient battles still circulate in some indigenous populations today. In other words, the descendants of those who survived that ancient epidemic continue to carry biological evidence of resilience—genetic defenses shaped through centuries of selection. It is a sobering reminder that some of our most intimate evolutionary struggles have been not with other humans, but with the parasites that tried to kill us from within. These genetic and archaeological clues make one thing clear—parasites helped shape us into who we are. They not only influenced our biology but also journeyed with us as we spread along the globe, determining where and how human societies would flourish.

CARRIED ALONG: PARASITES ON THE MOVE WITH HUMAN MIGRATIONS

Human history is a story of movement—across continents, through climates, over centuries. But wherever humans went, they were never alone. Lurking in their guts, clinging to their clothes, or coursing silently through their bloodstreams, parasites traveled alongside them. These unseen companions found new ecosystems, new hosts, and sometimes tragic new consequences. Tracking the journey of parasites is, in many ways, tracking the journey of humans themselves. Even in our deep prehistory, parasites offer clues to our wanderings. The types of lice that early humans carried out of Africa, for instance, reflect changes in our clothing, social structure, and migration into cooler climates.[5] When we invented garments, we did not just protect ourselves from the elements; we created new habitats for lice, which hitched rides to far-flung environments. Archaeologists have also discovered eggs of intestinal parasites in fossilized feces, revealing that wherever humans migrated, their worms came too.[6] In chapter 8, I discuss the fascinating field of paleoparasitology in more detail. What is important to note is that these ancient infestations underscore how parasites have been our traveling partners since the very beginning.

Sometimes these parasites found paradise in new lands. As Austronesian seafarers pushed into the Pacific, they brought rats, fleas, and mosquito-borne diseases across the ocean. But these parasites could thrive only if local conditions, like temperature, hosts, and vectors, were just right. Islands without the necessary snails, insects, or mammals remained parasite-free. Others, with the full ecological cast, became new strongholds for disease. The success or failure of a parasite's journey often depended on this hidden ecological calculus.

When human networks expanded into full-scale trade, the stakes of parasite transport escalated dramatically. Along the Silk Roads, traders carried silk, spices—and death. One of the most infamous stowaways was of course *Yersinia pestis*, the bacterium responsible for the Black Death. Likely spread by flea-infested rodents, this microbe traveled thousands of miles from Central Asia to Europe, igniting a pandemic that killed as many as two hundred million people.[7] This fourteenth-century plague was not just a medical crisis—it was a reshuffling of societies, economies, and belief systems, all unleashed by a microscopic traveler in a world without airplanes.

The age of transoceanic exploration brought perhaps the most devastating biological exchange in human history. As Europeans set sail to the Americas, they unleashed a storm of pathogens. Smallpox, measles, typhus, and influenza tore through Indigenous populations with catastrophic speed. But it was not only viruses that came ashore. Eukaryotic parasites joined the invasion. *Plasmodium* species and various intestinal protozoans were all part of this deadly caravan. With no prior exposure or immunity, many Native American communities suffered mortality rates of 70 to 90 percent within decades.[8] Entire civilizations unraveled, not through conquest, but through contagion. The resulting depopulation was so extreme that it helped seed the myth of the "virgin wilderness"—an empty land waiting to be claimed, when in truth it had been emptied by epidemic disease.

But disease flowed both ways. While scholars still debate whether syphilis made its way back to Europe from the Americas, other parasites certainly did.[9] The transatlantic slave trade added new layers to this global reshuffling. Africans forcibly brought to the Americas carried with them not only trauma and labor but also pathogens adapted to tropical environments. Yellow fever, borne by African mosquitoes, arrived in the New World in the 1500s.[10] Malaria soon followed, establishing itself in the American South and Caribbean.[11] Eukaryotic parasites like *Leishmania*, which now exist in multiple species and subspecies across South America, also became part of this biological exchange. These introduced and emerging diseases reshaped colonization patterns. Some tropical regions became virtual death traps for European settlers, while others became reliant on enslaved laborers who were more resistant to local parasites.

Not every parasite journey ended in success. Environmental mismatches often broke the chain of transmission. Cold Canadian winters, for example, prevented many tropical parasites from taking root. Likewise, European livestock frequently failed to thrive in parts of Africa, where they encountered unfamiliar pathogens and relentless biting flies. These ecological dead ends remind us that while humans could migrate nearly anywhere, some of their parasites were far more selective—demanding precise conditions to survive and reproduce. In chapter 1, I discussed the enemy-release hypothesis, which states that when a species is introduced to a new environment, it may leave its parasites behind. In its native range, the species is kept in check by natural enemies, especially parasites, which suppress its growth, reproduction, or survival. But when it arrives in a new region where its parasites cannot follow—perhaps due to the absence of a required insect vector or secondary host—it escapes those biological checks and balances. Freed from its enemies, the species can grow faster, live longer, and reproduce more prolifically than it ever could back home. It thrives simply because it has ditched its parasites.

Time and time again, these conditions were met. With every wave of migration and trade, parasites swept across the world—not as footnotes to history, but as potent, if invisible, forces that helped shape it.

WAR AND REROUTING THE COURSE OF CIVILIZATIONS

In the early 1300s, Italian city-states like Venice and Florence stood at the height of their cultural and economic influence. Florence was a financial titan, its powerful banking families—like the Bardis and

Medicis—extending credit across Europe. Its textile industry was booming, making it one of the wealthiest cities in Christendom. Venice, meanwhile, dominated maritime trade in the Mediterranean. With its formidable merchant fleet and a stable oligarchic government, it served as a vital conduit between East and West.

Now imagine you are a young, ambitious man from a small village in southern France in the year 1347. Born to peasant stock but gifted with skill and drive, you set out to forge a better life. Your goal? To reach one of the great northern Italian cities—perhaps to join a merchant guild or sign onto a ship headed for distant shores. You trek eastward, crossing the Alps into the urban bustle of Milan or Turin, where you find short-term work. But something is wrong. People are falling ill—quickly and in great numbers. In the streets, whispers swirl of a mysterious sickness spreading from the ports. Still determined, you press on toward Venice, hitching a ride with a merchant caravan. But as you arrive, the city is descending into chaos. The port smells of rot. Corpses line alleyways. Monks and physicians have fled or fallen ill themselves. The hospitals are full—turned into mass graves—and law and order have broken down. You have arrived not at the doorstep of opportunity, but at the epicenter of one of history's most devastating pandemics.

The Black Death has begun.

Between 1347 and 1351, the bubonic plague tore across Europe with astonishing speed, carried by rats and fleas aboard merchant ships and caravans. By the time it ran its course, it had killed an estimated 30 to 60 percent of Europe's population. Venice, Florence, Paris, and London—none were spared. The plague did not just claim lives; it upended the structure of society. Labor shortages toppled feudal systems. Religious institutions faltered. Art and literature took a dark, introspective turn.

For modern readers, the closest parallel might be the early days of the COVID-19 pandemic in 2020. ICU nurses in cities like

New York witnessed firsthand the overwhelming toll of an invisible pathogen. But even then, we had a vast scientific arsenal: ventilators, PPE, genome sequencing, and eventually vaccines. In the fourteenth century, there were no microscopes, no germ theory, no understanding of how disease spread. It would be another two centuries before Antonie van Leeuwenhoek, peering through a handcrafted lens, became the first human to glimpse a microbe.

As humans moved, through migration, trade, and conquest, parasites and pathogens traveled with them. Sometimes they lurked quietly in the background. Other times, they erupted into pandemics that did not just claim lives but reshaped history. From the bubonic plague to tropical fevers, infectious diseases have served as invisible agents of change, toppling empires, redirecting economies, and transforming societies. In this section, I highlight a few pivotal episodes when parasites altered the course of history—not with swords or treaties, but with fevers, flies, and the weight of microscopic inevitability.

In 1802, Napoleon Bonaparte sent a powerful French force to Saint-Domingue (now Haiti) to crush a slave rebellion and reassert imperial control. But the campaign unraveled—not because of the rebels' gunfire, but from the bite of a mosquito. Yellow fever, a virus transmitted by the mosquito species *Aedes aegypti*, swept through the French ranks with staggering speed.[12] Roughly four-fifths of Napoleon's army perished. Even the wife of the French commander, Leclerc, succumbed. The virus decimated entire regiments, leaving surviving troops too sick to fight. What the revolutionaries could not accomplish alone, a mosquito-borne disease did. Napoleon's dreams of a western empire collapsed, and in the aftermath, France sold the vast Louisiana Territory to the United States. A microscopic parasite in the Caribbean did not just help secure Haitian independence; it redrew the map of North America.

Haiti was not an anomaly. Time and again, parasites sabotaged colonial ambition. In the eighteenth and nineteenth centuries, large swaths of tropical Africa were dubbed "the White Man's Grave" because of sky-high European mortality from malaria and yellow fever. Before the discovery of the anti-parasitic drug quinine, European settlers and soldiers rarely survived more than a few months in these environments. In many ways, tropical parasites served as a hidden line of defense for Indigenous populations. Even in the twentieth century, parasites continued to frustrate imperial projects. During the construction of the Panama Canal, tens of thousands of workers died from mosquito-borne diseases before large-scale vector control efforts turned the tide.[13] These pathogens acted as equalizers in otherwise asymmetric power struggles, crippling colonial powers while sparing populations with generations of acquired immunity. In fact, through much of military history, more soldiers died from disease than from direct combat.[14]

During World War II, the Pacific Theater saw malaria rates exceed 80 percent among U.S. Marines stationed in Guadalcanal.[15] Across the theater, large infantry divisions were decimated not by Japanese bullets but by fever-ridden jungle mosquitoes carrying a deadly parasite. Treatment options were limited—quinine supplies had been cut off by Japanese control of Southeast Asia, and atabrine, a synthetic substitute, was poorly tolerated and mistrusted by troops (some falsely believed it caused impotence), leading to poor compliance.[16] Commanders quickly realized that they were fighting two wars: one against an army adversary, and another against a parasite. The crisis spurred one of the largest military public health mobilizations in history. The U.S. military invested heavily in entomology, vector control, and parasitology research. Engineers drained swamps and deployed DDT to kill mosquito larvae; medics distributed atabrine under strict orders; researchers developed new antimalarial drugs like chloroquine and mepacrine. While malaria was the most immediate threat, soldiers in the Pacific Theater also encountered other parasitic adversaries—including filarial nematodes. *Wucheria bancrofti*, the causative agent of lymphatic filariasis, was endemic to many tropical island regions where Allied troops were stationed. In some areas, infection rates were extremely high among locals, and soldiers deployed for long periods were at high risk.[17] The parasite uses night-biting mosquitoes as its vector. Once inside the human host, the worms migrate to the lymphatic system, where they block or damage lymphatic vessels, which are responsible for maintaining fluid balance, immune surveillance, and fat absorption. While not as grotesque as its cousin the Guinea worm (*Dracunculus medinensis*), the resulting obstruction can cause debilitating swelling of the limbs and genitals, a condition known as lymphedema or elephantiasis. Some servicemen developed chronic disability that persisted long after the war ended. At the time, treatment options were limited, and vector control strategies were still in their infancy. Though less lethal than malaria, lymphatic filariasis had a profound impact on troop readiness and illustrated a critical truth of tropical warfare: In parasite-rich environments, even victory can come at a biological cost.

The lessons learned shaped tropical modern medicine and vector control strategies in peacetime settings. World War II was not just a battle of nations; it was a proving ground for the science of parasitology. The war catalyzed innovations in diagnosis, treatment, and vector control that saved lives far beyond the battlefield. It also reminded public health officials that in the right ecological context, parasites can become strategic adversaries, undermining even the most technologically advanced militaries. These experiences helped lay the groundwork for what would later evolve into the One Health concept—the recognition that human health, animal health, and environmental conditions are inextricably linked.[18] Military medicine in the tropics revealed that to control disease, it was not enough to treat the human host; one had to understand the transmission ecology—the parasite, the vector, the reservoir species, and the habitat they all shared.

Not all parasitic influences were as explosive or immediate. Some worked slowly—gnawing at health, productivity, and social stability over centuries. In ancient agricultural societies like Egypt and China, diseases such as schistosomiasis (snail fever) likely suppressed economic output and even shaped religious practices as communities sought supernatural explanations for chronic suffering.[19] Modern archaeological research confirms this: In ancient Nubia, around 1,500 years ago, the adoption of irrigation created stagnant water, prime habitat for freshwater snails, the intermediate hosts for *Schistosoma* worms. Irrigation also opened up new ecological niches, spreading these snails and expanding the parasite's range. In one Nubian fossil-tissue study, researchers found about 25 percent of individuals from a canal-irrigated site carried *S. manosni* antigens, compared with about 9 percent from a less irrigated site—a telling hint that irrigation practices amplified parasitic burden.[20] Here, human innovation fed a parasite, which in turn burdened the population—an unintended trade-off written in their bodies.

Other water-linked parasites followed similar paths. In many parts of Africa and Asia, the Guinea worm tormented rural communities for millennia. As discussed in chapter 1, this parasite is transmitted by drinking water contaminated with infected copepods. The worm causes excruciating pain as it emerges slowly from the skin, often incapacitating individuals for weeks during critical agricultural periods. Likewise, *Onchocerca volvulus*, the parasitic worm behind river blindness, has long been endemic in fast-flowing river regions where blackflies (the vectors) breed. While rarely fatal, it causes intense skin irritation, permanent vision loss, and social stigma—particularly devastating in rural farming communities where labor and eyesight are indispensable. Both parasites did not just harm individuals; they eroded the economic and social fabric of entire regions.

Scholars increasingly argue that chronic parasitism—from malaria, helminths, and waterborne pathogens—helped trap certain regions in long-term cycles of poverty and underdevelopment. Meanwhile, societies that escaped or controlled these burdens—by climate, sanitation, or sheer geographic luck—were more likely to flourish. In this way, some of humanity's oldest enemies did not just shape bodies; they shaped civilizations.

For those linguistically inclined, it is no coincidence that one of history's most devastating parasitic diseases, malaria, derives its name from the Italian words *mal* ("bad") and *aria* ("air"). Long before germ theory, people sensed that something in the fetid air of swamps and lowlands made them sick. They were wrong about the medium—but right about the message. Parasites often emerged from the environments we built or settled in, following us into fields, canals, cities, and trade routes.

By now it is clear that parasites have drastically affected human survival and societal trajectories. But might they have also shaped our most intimate belief systems—our religion?

PARASITES IN MYTH AND RELIGION: INVISIBLE PUNISHMENTS AND SACRED LESSONS

When historians speak of parasites shaping human destiny, they often point to genetics, trade, or war. But another realm of influence lies hidden in plain sight: religion and myth. Parasites were such intimate companions of early humanity—sapping strength, disfiguring bodies, and erupting from skin—that it is no wonder they became woven into our deepest cosmologies. In sacred texts and folklore, worms and plagues are never just biological afflictions; they are divine messages, punishments, or metaphors for sin. Parasites were not only in our blood and bellies; they were in our stories.

Take the Bible. The Old Testament is steeped in pestilence; boils, worms, and infestation are often presented as instruments of divine wrath. One of the most sensational examples comes in the Book of Acts when King Herod Agrippa I, after accepting praise as though he were divine, "was eaten by worms and died."[21] The historian Josephus gives a parallel account in *Antiquities*, placing his death in Caesarea following a protracted illness and acute pain. Some medical historians have proposed that the description—especially the abdominal agony and the "worms" metaphor—might hint at a parasitic disease such as schistosomal damage, though the diagnosis cannot be confirmed with certainty.[22]

The Book of Job, another powerful theological reflection on suffering, offers visceral imagery of parasitic decay: "My flesh is clothed with worms and clods of dust; my skin is broken and festering." Here, worms function as both physical torment and metaphors for existential despair, linking parasite imagery to spiritual ruin. Prophetic texts often use worm-infested decay as metaphors for moral corruption. After pronouncing judgment on proud and wayward rulers, Isaiah declares: "Maggots are spread out beneath you, and worms cover you." Later he laments: "Their worm shall not die, neither shall their fire be quenched." In these passages, worms symbolize unrelenting spiritual decay, both a warning and a symbol of persistent sin. These biblical expressions are not merely metaphorical; archaeological science adds somber weight to them. At Armon HaNatziv in Jerusalem, excavations beneath a 2,700-year-old stone toilet revealed more than 850 fossilized parasite eggs, including whipworms and tapeworms, from affluent First Temple–period residents, demonstrating that parasites plagued even the elite.[23]

In the Hindu medical tradition of Ayurveda, internal parasites were not an afterthought but part of a rich disease theory. The Sanskrit term *krimi* encompassed worms, microbes, and invaders of the body; texts describe them as agents of weakness, digestive distress, and emaciation.[24] Ayurvedic treatments for krimi often combined herbal formulas with ritual cleansing and purification—an implicit recognition that illness could be both bodily and spiritual. While religious fasting is broadly present in Hinduism, Buddhism, Christianity, and Islam—often understood as spiritual purification or self-discipline—it may also echo an older intuition that withholding food can help the body resist internal invaders. In this sense, parasites blur the boundary between the material and the metaphysical; they are both physical intruders and symbols of spiritual imbalance.

Food taboos in Abrahamic religions may also have biological shadows. The ban on pork in Jewish kashrut and Islamic halal codes is often seen as spiritual discipline, but parasitologists have long noted a more visceral logic: Pigs harbor *Trichinella spiralis* and *Taenia solium*, parasites that in premodern societies made eating undercooked pork a gamble with trichinosis or tapeworm disease. In such conditions, divine law functioned as inadvertent public health.[25] Generations of suffering from unseen worms may have etched the commandment into cultural memory. The avoidance of shellfish carries similar echoes: Bivalves can host pathogens such as *Vibrio* bacteria, dangerous to humans when eaten raw. Sacred rules thus helped define what was safe to eat and, in doing so, which parasites would thrive, or vanish, from human life.

Other traditions made parasites central to myth. In parts of sub-Saharan Africa, the long emerging, excruciatingly painful worm associated with Guinea worm disease was often interpreted not simply as illness but as a curse or spiritual trial. Villagers attributed its emergence to witchcraft or ancestral malice; healing included ritual removal and appeasement.[26] In Andean folklore, while direct references to intestinal worms are rarer in the literature, creatures of decay and internal torment are common; it is plausible that parasites once played a role in such stories. Across cultures, parasites blurred the boundary between biology and demonology: What could not be seen but caused deep suffering was easily cast as supernatural.

Parasites also became metaphors for corruption and greed in religious and philosophical writing. Early Christian sermons likened sin to a worm that eats from within, unseen until it is too late. Medieval European texts described moneylenders and corrupt officials as "parasites"—a term that began as a neutral Greek word for "one who eats at another's table" but over centuries absorbed the biological connotation of consumption without reciprocity. In Dante's *Inferno*, sinners are tormented by grotesque afflictions that echo real parasitic diseases: eternal itching, bloating, gnawing from within. The imagery was not abstract; it was drawn from lived experience of worms, lice, and fleas that stalked daily life. Islamic scholarship, too, recognized the moral weight of parasites. The ninth-century physician-philosopher al-Razi wrote about intestinal worms as serious ailments in the Islamic Golden Age. Cleanliness—ritual washing before prayer, dietary restrictions—was more than spiritual discipline; it was parasite control. Religious practice encoded survival strategies in ways believers might never have recognized consciously but that helped keep communities healthier.

Even in modern times, parasites retain this symbolic power. When Karl Marx denounced capitalism as a system on which the bourgeoisie "sucks the blood" of the working class, he drew directly on the imagery of leeches and worms. In literature, from Bram Stoker's *Dracula* to the body-horror films of David Cronenberg, parasitism has remained a metaphor for domination, exploitation, and loss of autonomy. The religious imagery of worms as sin and demons found new life in political critique and art.

What unites these traditions is the recognition that parasites, invisible yet devastating, demanded cultural interpretation. Unlike a tiger or a flood, they could not be seen or fought head-on. They were intimate invaders, part of the body yet not of the body. That strangeness invited myth. And in myth, parasites did more than frighten; they taught. They warned against gluttony, impurity, and disobedience.

They offered explanations for suffering that was otherwise inexplicable. From the plagues of Egypt to the dietary codes of the Torah, from Ayurvedic texts to Andean legends, parasites may also have been the silent players. They are one of the hidden actors behind ritual cleansing, fasting, food taboos, and moral metaphors. To write them back into the story is not to diminish the spiritual significance of those traditions but to see more clearly how our biology and belief have always been intertwined.

Parasites shaped our genomes, our societies, and our cities—but they may have also shaped our sacred imaginations. They were not only in our blood; they were in our prayers. Yet if parasites haunted the myths of our ancestors, they also pressed their weight on something even more intimate: the very way we think. What if the influence of parasites goes beyond scars on the body and symbols in scripture—what if they have touched the architecture of human intelligence itself? It is a provocative, controversial idea, but one that deserves coverage because of its far-reaching implications.

CULTURE UNDER THE MICROSCOPE: PARASITES AND INTELLIGENCE

Parasites have shaped not only our biology and our belief systems, but perhaps even the way human societies think and organize themselves. There is a highly controversial idea that has been floating around parasitology and evolutionary psychology for decades, and though it has always been contentious, it deserves attention because of the questions it raises. It is the hypothesis that widespread parasitic infection, especially in early childhood, can depress cognitive development at the population level, potentially influencing everything from school performance to the average IQ scores of entire nations.

The origins of this idea reach back to early twenthieth-century tropical medicine, when colonial doctors working in Africa and South America observed that heavily parasitized populations often showed fatigue, lethargy, or underperformance in school. Their interpretations were steeped in bias and racism, but behind them lay an empirical observation: Chronic infection with worms, protozoa, and blood parasites can rob children of energy, nutrients, and developmental time. By the late twentieth century, Corey Fincher and Randy Thornhill reframed this as the parasite-stress hypothesis, which we briefly touched on earlier in the book. Their argument was that parasites do not just sap individual health—they may also shape the behavior and capacities of entire societies, fostering collectivist cultures, stronger in-group biases, and, in the most extreme versions of the theory, influencing cognitive performance across populations.

How might parasites alter intelligence? The pathways are multiple. Chronic worm infections, such as hookworm, whipworm (*Trichuris*), and roundworm (*Ascaris*), bleed the body of iron and protein, leading to anemia and stunted growth. Malaria and schistosomiasis inflame the immune system, pulling metabolic energy away from brain growth and learning. Maternal infections compromise fetal development, reducing oxygen and nutrient flow to the fetus. And in daily life, infected children often miss school or, when present, sit at their desks exhausted, unable to concentrate. The assault is not just on biology but on opportunity: Parasites steal time, focus, and the chance to learn.

Large-scale studies have attempted to measure this burden. In southeastern Nigeria, nearly half of secondary school students were infected with intestinal worms, most commonly *Ascaris lumbricoides*. Students with infections were far more likely to score in the "extremely low IQ" range, while uninfected peers performed at average levels. Nutritional deficits compounded these effects, underscoring the dual burden of malnutrition and parasitic diseases on cognitive development.[27] In Nepal, researchers showed that undernourished adolescents had a dramatically higher likelihood of scoring below 70 on IQ tests, even though no direct correlation with *Ascaris* was found—suggesting a tangled web of nutrition, infection, and poverty.[28] Other global analyses have noted strong correlations between parasite prevalence and national IQ averages. Even after adjusting for GDP, education, and nutrition, nations with higher parasite burdens tended to have lower average IQ scores.[29]

Still, correlation is not causation—and critics have been vocal.[30] The notion that national intelligence differences stem from parasite burden gained attention after a 2010 study by Eppig, Fincher, and Thornhill claimed a strong global correlation between infection rates and average IQ; but in 2024 that paper was formally retracted after editors concluded the underlying IQ datasets were deeply flawed and unreliable, illustrating how fragile such sweeping cross-national comparisons can be.[31] Beyond data disputes lies a deeper issue: IQ itself is a culturally loaded and contested measure. More importantly, regions with high parasite prevalence are also those facing poverty, malnutrition, and weak infrastructure. Are parasites the cause of lower performance, or simply one thread in a larger web of structural disadvantage? Many scholars argue the latter. Yet field evidence from randomized interventions tells a more grounded story: Deworming, nutritional supplementation, and malaria control consistently improve school attendance, attention, and even long-term earnings. Parasites may not determine destiny, but their removal clearly helps.

Some of the most striking evidence comes from natural experiments in deworming. In Kenya, a landmark randomized trial led by Edward Miguel and Michael Kremer in 2004 provided mass deworming to schoolchildren. Attendance rose by 25 percent, spillover benefits extended to untreated children, and follow-up studies found higher rates of secondary school completion and better wages in adulthood.[32] Similar, though sometimes more modest, effects have been documented in India and other high-burden regions. These results have made deworming one of the most cost-effective global health interventions ever proposed, championed by NGOs and the World Bank alike. The idea is simple but profound: Fighting parasites can yield not just healthier bodies but stronger economies and sharper minds.

The debate continues. Follow-up meta-analyses have suggested the effects are smaller than early advocates claimed, and some trials have failed to replicate dramatic improvements.[33] The controversy illustrates the difficulty of untangling parasites from the wider environment of poverty. Yet even if the impact is less sweeping than first imagined, it is hard to deny that reducing parasite burden creates conditions more favorable to learning, growth, and opportunity.

What makes this debate fascinating is not only its scientific stakes but its philosophical ones. If parasites have indeed blunted cognitive development for generations, then they have shaped not just biology, but history. Civilizations struggling under heavy parasitic loads may have faced hidden ceilings—not from a lack of ingenuity or willpower, but from infections that stole energy and focus in childhood. Conversely, societies that escaped these burdens, whether by geography, climate, or luck, may have gained an invisible head start. Parasites, in this light, become more than pathogens. They become geopolitical actors, silent but consequential in the destinies of nations. The policy implications are enormous. Parasite control is more than a medical intervention; it is an investment in cognitive infrastructure. It is an investment in human potential, as vital as schools, roads, or internet access. Deworming campaigns, vector control, and sanitation systems may yield dividends not just in fewer sick days but in more inventors, leaders, and citizens able to fulfill their potential. The question is whether we are ready to reframe these interventions as engines of empowerment rather than mere disease prevention.

Once again, we are left with the insight of William Campbell, the Nobel-winning parasitologist: "It is time for parasites to get a little more respect." Respect, in this case, does not mean admiration, but recognition of their power. If parasites can depress the collective intelligence of people, then eradicating them may indeed unleash the opposite: a renaissance of minds long dulled by unseen burdens. Parasite control, seen this way, is not merely a health intervention but a strategy for unlocking human potential on a global scale.

And yet, respect also means caution. If parasites have shaped our genomes, our cultures, and even our civilizations, then what are the consequences of erasing them entirely? Could some parasites play hidden roles we do not yet understand? Could their disappearance ripple through ecosystems in unexpected ways? These are not just scientific questions but ethical ones, and they point toward a deeper debate, one we will visit in more detail in the following chapter.

PARASITES AS CATALYSTS FOR TECHNOLOGICAL AND URBAN INNOVATION

Finally, let us end this chapter on a high note. Most stories about parasites focus on what they take: our nutrients, our energy, our health. But what if we have been telling only half the story? What if, in their relentless assault on the body, parasites have also given us something in return—not by design, of course, but as a kind of evolutionary provocation, forcing us to innovate, organize, and build in ways we might not have otherwise imagined? Consider the American South in the early 1900s. At the time, much of the region was ravaged by hookworm disease—a soil-transmitted nematode parasite that enters the body through bare feet and slowly drains its host of energy. Victims were anemic, fatigued, and often dismissed as lazy or simple. This was not just a medical issue; it was a cultural one. An entire region was being pathologized.

Enter the Rockefeller Sanitary Commission, a privately funded campaign launched in 1909 to eradicate hookworm across the southeastern United States. What began as a targeted effort to deworm the rural South soon evolved into something far greater: one of the first large-scale public health initiatives in American history. It deployed physicians, mobile labs, and health educators across eleven states, testing more than a million people and treating hundreds of thousands. Just as crucially, it introduced sanitation infrastructure—latrines, screened-in privies, public health messaging—and helped normalize the idea that governments could, and should, take an active role in preventing disease. In a strange twist of fate, the parasite became a kind of public health muse. It catalyzed the creation of rural health departments, legitimized public health as a profession, and laid the groundwork for future efforts against malaria, tuberculosis, and more.[34] The hookworm did not just take—it pushed us to build.

This was not a one-off. Throughout history, parasites and the diseases they carry have been behind some of our most transformative developments. Let us take a closer look at some of these.

When we think of ancient Rome, we often imagine engineers building monumental aqueducts and grand public latrines, symbols of civilization and hygiene. Indeed, the Romans constructed more than eleven aqueducts, supplying water across the empire, and engineered multi-seat public toilets connected to the expansive Cloaca Maxima sewer system. These systems aimed to whisk away waste and replenish fountains and baths with fresh water. The intent was clear: to rid cities of stench and disease by sweeping away filth and bringing in clean water.

Yet modern parasitological studies have revealed a surprising twist. Despite these advanced sanitation systems, Roman cities remained infested with parasites like whipworm, roundworm, amoebic dysentery, and fish tapeworms—often at rates as high or higher than in later medieval or even less urbanized societies.[35] Far from providing relief, communal baths and reused sponges may have spread infections, and human waste from latrines was reused as fertilizer, amplifying exposure. Rome's sanitation may have masked smells, but it did not curb the microscopic scourge beneath. So why did Romans build these monumental sanitation systems? Because even without germ theory, they knew filth was linked to illness. They did not realize their interventions were incomplete, much like the early twentieth-century hookworm efforts. But their ambition laid essential groundwork. The aqueducts and sewers were not cures, but these early attempts at environmental engineering would, millennia later, blossom into modern public health infrastructure.

During the COVID pandemic that began in 2020, an entire generation of Americans experienced large-scale quarantine for the first time. Public health officials had long considered such measures a last line of defense. In 2007, a man infected with extensively drug-resistant tuberculosis (XDR-TB) was forcibly isolated in a CDC facility, marking the first federal quarantine order in more than forty years.[36] Sporadic regional quarantines had been implemented in the twenty-first century as well: in 2003, SARS outbreaks in parts of Canada and China led to the isolation of more than 25,000 people.[37] In Toronto, hospitals closed to nonessential visitors, and health-care workers were told to stay home. In Beijing and Guangdong, buildings were sealed and public transportation suspended. The 2014–2016 Ebola outbreak in West Africa prompted the sealing off of entire districts in Sierra Leone, Liberia, and Guinea, with nationwide lockdowns implemented to find hidden cases.[38] A century earlier, in 1918, the Spanish flu prompted many U.S. cities to close schools, theaters, and churches, enforce mask mandates, and restrict movement; St. Louis famously acted early, resulting in far lower death rates than cities like Philadelphia.[39]

The practice of quarantine is neither a new concept nor, as some would like to believe, a smokescreen for curtailing individual freedoms and intentionally causing economic disasters. It is a time-tested idea that can trace its roots all the way back to fourteenth-century Venice. The word itself comes from the Italian *quaranta giorni* ("forty days"), referring to ships waiting offshore before docking during plague outbreaks. During the plague, entire quarantine islands, such as Lazzaretto Nuovo, were established as zones for the sick and their goods, slowing the spread of infection. These measures, far from being primitive panic responses, were early expressions of a society trying to understand and contain the unseen forces shaping human life. They mark the birth of epidemiological thinking—the attempt to create buffer zones not just between the sick and the well, but between human civilization and the microbial world.

If quarantines were about containment, microscopy was about revelation. By the seventeenth century, another revolution was quietly brewing—not in policy, but in perception. The Dutch lens grinder and amateur scientist Antonie van Leeuwenhoek peered into a handcrafted microscope and reported a stunning discovery: tiny, motile creatures teeming in rainwater, saliva, and even his own feces. He called them *animalcules*. What van Leeuwenhoek did not know—what no one could fully grasp at the time—was that many of these organisms were parasites. He had, quite literally, gazed into the unseen world that quarantine had been trying to keep at bay. Though van Leeuwenhoek's motivations were mostly observational, his discoveries ignited a centuries-long effort to connect these microscopic life-forms with disease. As microscopy improved, the nineteenth century saw it transformed from a curiosity into an indispensable tool of public health. Louis Pasteur and Robert Koch, two of the founding figures of microbiology, used microscopes to identify and isolate the bacteria behind anthrax, cholera, and tuberculosis—pathogens that had long been suspected but never seen. Around the same time, the first microscopic descriptions of parasitic organisms began to appear in the scientific literature: *Plasmodium* in the blood of malaria patients, *Trypanosoma* wriggling in the bloodstreams of sleeping sickness victims, and *Leishmania* hiding in swollen lymph nodes. Entire fields—parasitology, protozoology, epidemiology—were born from the microscope's eye. This was not just a scientific triumph; it was a radical reordering of the way people understood health, illness, and the inside world around them. What had once been blamed on curses, bad air, or divine wrath could now be traced to tiny organisms swimming just out of sight. For the first time, the enemy had a face.

PUTTING IT ALL TOGETHER

Throughout this chapter, we have peeled back the curtain on one of the most overlooked forces in human history: the parasite. Far from being mere footnotes in our evolutionary story, parasites have acted as agents of natural selection, catalysts of public health reform, arbiters of war and empire, and even, perhaps, hidden influencers of intelligence and innovation. We have seen how ancient parasites sculpted our genomes, embedding defensive mutations into our DNA—sometimes beneficial, sometimes burdensome. From sickle cell disease and G6PD deficiency in malaria-endemic regions to immune-modulating adaptations in the Amazon against Chagas disease, our genetic code is a palimpsest of parasitic pressure past. We have followed these pathogens across space and time, tracking their migration alongside ours—from lice clinging to the clothes of Paleolithic humans to the protozoans that arrived with European conquest and rewrote the demographic fate of the Americas.

We have also explored how parasites shape not just individuals but societies. Parasites altered trade routes, halted empires, and undermined colonial campaigns. They reconfigured labor systems, urban sanitation, and might even have played a subtle role in religious belief. Quarantine, vector control, and public health outreach emerged not from theoretical planning but from the gritty realities of disease management—born of desperation as much as insight. In this light, parasites can be seen not merely as destroyers but also as provocateurs of ingenuity. From the invention of microscopy to the architecture of modern sewage systems, many of our greatest innovations arose from our need to confront them.

Parasites continue to influence us. The hypothesis that they suppress cognitive development and national IQ, while controversial, raises profound questions about the long-term effects of chronic infection. If parasite eradication can unlock developmental potential, then controlling parasites is not just a medical mission but a geopolitical strategy. It is a lever for empowerment and equality.

8

THE GHOSTS OF EVOLUTION PAST

"Do you think life is nothing but a fragile, thin, soft shell clinging to the surface of this planet?"

"Isn't it?"

"Only if you neglect the power of time. If a colony of ants continue to move clods the size of grains of rice, they could remove all of Mount Tai in a billion years. As long as you give it enough time, life is stronger than metal and stone, more powerful than typhoons and volcanoes."

—Cixin Liu, *Death's End*

Aside from being the only planet we know that harbors life, Earth is, on a cosmic scale, nothing special. In the vastness of the universe, it is a dust mote suspended in the darkness, orbiting a modest star on the fringe of a common galaxy. On other worlds, perhaps there are storms greater than our continents, tidal waves that reshape entire landscapes, and maybe—just maybe—life. Some of that life may have endured for thousands, even millions, of years, only to vanish—the blink of an eye in the timeless machinery of the cosmos. On Earth, too, life rises and fades. Species are born, live briefly, and then disappear. For the fortunate ones, the ones that inherit the right suite of genes, there is a chance, however slim, to pass those genes on. In a universe ruled by entropy, heredity is the closest thing we have to immortality.

Extinction is not the exception; it is the rule. The fate of most life-forms—past, present, and future—is erasure. On the evolutionary trees we draw, we rarely show the true shape of life's history. A true evolutionary tree would resemble a forest in perpetual autumn: branches ending abruptly while only a few shoots continue upward. Most lineages have vanished long before the present. Which of those branches will persist into the future is anyone's guess. For a brief moment, an evolutionary heartbeat, they are here—and then they are not. This impermanence is perhaps the only constant in biology, an inescapable shadow cast across the arc of every lineage.

Paradoxically, it is impermanence that makes evolution possible. It clears the stage and opens new doors. When the dinosaurs were swept away sixty-five million years ago, it was not just their bones that fell silent; their parasites vanished too. Although most parasites do not fossilize, some do. In ancient amber, entombed like relics in time, we find their shadows: blood-feeding insects, protozoan spores, and parasitic worms that once clung to reptilian hosts now long gone. Their extinction was swift and absolute—not just the *T. rex*, but the lice that fed on its scales; not just the herbivorous hadrosaurs, but the intestinal nematodes that coiled through their guts.

These ghosts, preserved in stone or resin, offer something rare: a glimpse of the hidden passengers of evolution, the ones that lived in the margins, the folds, the shadows. They remind us that when a host goes extinct, it often takes an entire unseen ecosystem with it.

A VANISHING LEGACY

Imagine this scenario: A single bird species, endemic to a remote island, is teetering on the edge of extinction. Conservationists monitor the final handful of individuals with desperate attention. They track their calls, guard their nests, and ration the remaining habitat as if time could be stalled. Then, one season, there are no eggs. The last male continues to live for a few years, in a state of perpetual loneliness, and finally sings into silence. When he dies, headlines mourn the loss of a unique lineage, a branch snapped from the tree of life.

But few, if any, mourn what died with him.

In the feathers of that bird lived a species of louse found nowhere else on Earth. In its gut was a helminth that had co-evolved with it over millennia, in its blood a microscopic parasite that had adapted exquisitely to its physiology and immune system. When the bird vanished, so did they—no names, no headlines, no fossil record, a quiet extinction within an extinction, a shadow erased before we even knew it was there.

We tend to think of extinction as the loss of a single species, the last individual fading out. But in reality, it is less like the loss of a single species than the collapse of a cathedral, taking with it the intricate structure it once supported. Parasites often occupy these hidden chambers. They are the forgotten passengers of evolution, vanishing not in isolation but in tandem with their hosts. In that sense, extinction means the complete removal from Earth's gene pool of entire evolutionary histories, many of which we never even knew existed.

In the grand narratives of deep time, we elevate the dinosaurs, the mammoths, the saber-toothed tigers, all of what zoologists call the "charismatic megafauna"—and for good reason. They are fascinating creatures that have inspired myths and legends; their faces have adorned nonprofits that exist for truly noble purposes. Their bones fill museum halls, their fossils fetch tens of millions of dollars in high-profile auctions, and their extinction stories inspire documentaries and dioramas. But behind those towering skeletons were worlds within worlds: ticks in the fur, protozoans in the blood, tapeworms in the intestines. These parasitic hitchhikers rarely get a mention. Yet they may hold some of the most important clues to understanding how evolution actually works.

This is where a quiet, often overlooked field steps in: paleoparasitology—the study of ancient parasites. It is, by necessity, a science of scraps. Parasites rarely fossilize. Most are soft-bodied, small, and hidden deep within organisms. But sometimes they leave traces: a whipworm egg nestled in an eight-thousand-year-old coprolite (a more endearing term than "fossilized feces"); a tick trapped in amber, still clutching a fragment of reptilian skin; liver flukes embedded in fossilized lesions.[1] These remains are fleeting, partial, and fragmentary, but they are enough to tell us that parasites were there. They too had lineages; they too evolved, diversified, adapted, and died out. Our newfound ability to access ancient DNA (aDNA), as a result of the molecular revolution, has given paleoparasitology an unprecedented boost. With genetic tools, scientists can recover the molecular fingerprints of parasites from mummies, fossilized feces, and even the sediment of ancient lake beds. Suddenly, we can reconstruct parasite genomes from the Ice Age! We can trace lineages that long predate agriculture, or see how climate change altered the range of disease-causing organisms thousands of years ago. Yet, despite these revelations, parasites remain the footnotes of extinction—the afterthoughts. In conservation biology, they are often treated like baggage—problems to be eradicated, not species to be protected. In paleontology, they are buried beneath layers of more glamorous stories. But perhaps, if we learn to read those footnotes more carefully, we may uncover a deeper truth: that the story of life on Earth is not just about predators and prey, bones and teeth, but also about who lived in the margins. The intimate, parasitic, often invisible relationships that shaped evolution from within.

What we have missed in ignoring parasites is not just detail; it is meaning. Parasites are not evolutionary flukes (no pun intended); they are evolutionary forces. They do not just accompany life; they shape it, direct it, end it, and sometimes even preserve it in the fossil record. This chapter is a tribute to those ghosts, the ones evolution forgot or tried to bury. Through the tools of paleoparasitology, we are beginning to listen to their whispers. We are beginning to uncover not just how parasites lived but also how they died, and what their disappearance can teach us about our own fragile legacy.

THE SILENT KILLERS OF PREHISTORY: PARASITES AND MASS EXTINCTION

Extinction is rarely a single blow. It is often death by a thousand cuts: climate upheaval, shifting coastlines, starvation, habitat collapse. But there is another agent of annihilation—quieter, more insidious, and often invisible in the fossil record—that may have haunted species on their final days. Long viewed as evolutionary afterthoughts or hitchhikers on the broader story of life, parasites may have played far more consequential roles in the grand extinctions of the past. Their impact would have come not as fiery impact craters or towering lava flows, but as creeping, invisible killers—amplifiers of stress, transmitters of disease, and agents of immunological sabotage.

We do not typically think of parasites as drivers of extinction. They are rarely fossilized and seldom seen in the dramatic iconography of extinction narratives. Dinosaurs, mammoths, saber-toothed cats—these are the stars of prehistoric demise. But if we zoom in, beneath their skin, into their guts and their bloodstreams, we find a hidden ecological layer: tapeworms, protozoans, flukes, viruses, and bacteria, many of which lived intimately with their hosts. And sometimes, it was these tiny organisms that helped tip the scales toward extinction. One of the most provocative theories suggesting parasites as coauthors of extinction is the hyperdisease hypothesis proposed by Ross MacPhee and Preston Marx.[2] According to this idea, novel pathogens introduced by invading species, often humans and their commensals, could have decimated immunologically naive megafaunal populations. In this model, it was not overhunting or climate change alone that brought the Pleistocene giants to their knees, but infectious disease outbreaks with no historical precedent in the host populations.

Picture this: A new host enters a landscape. Along with it comes a tick-borne pathogen or intestinal parasite foreign to the native herbivores. The infection spreads quietly at first, undetectable in the bones it leaves behind. But with no prior exposure, these large mammals have no immune memory to defend themselves. Herds thin; birth rates drop; calves die young. Reproductive collapse follows. Within a few generations, a stable population crumbles.

While conclusive fossil evidence is difficult to obtain for such biological events, circumstantial support comes from the synchronicity between megafaunal declines and human arrival in several parts of the world. In Australia, for instance, many large-bodied marsupials vanished shortly after the arrival of humans some fifty thousand years ago.[3] Could novel zoonoses, perhaps brought over by early domesticates, have contributed to this cascade? Even in modern times, we have seen echoes of how devastating this kind of introduction can be. The arrival of avian malaria in Hawaii, transmitted by introduced mosquitoes, drove several species of native birds to extinction.[4] These birds, having evolved in an environment free from these parasites, had no defenses. A similar scenario, played out thousands of years earlier in the Pleistocene, would leave few traces beyond absence.

Parasites may not always be the executioners, but they are skilled at kicking species when they are down. Climate-induced stress, whether from food scarcity, habitat loss, or thermal extremes, can suppress animals' immune systems, and in that vulnerable window, parasites can seize the moment. This dynamic is not just theoretical. Several studies have found that climate-driven physiological stress, such as temperature extremes, significantly impairs immune function in vertebrate hosts. Under such conditions, even parasites that might normally be kept in check can flourish, turning what was once a balanced coexistence into a lethal overload. A grim feedback loop emerges: Warming temperature compromises immunity, parasites proliferate, and the weakened host population spirals further toward decline.

This co-stressor role of parasites was echoed in an influential synthesis written by the ecologist Sonia Altizer and her colleagues more than a decade ago.[5] Highlighting the growing body of evidence showing how climate change could disrupt host-parasite dynamics on a global scale, their study warned that warming global temperatures can increase transmission rates, prolong seasonal windows of infection, and expand the geographic ranges of parasites into naive host populations. Such shifts, they argued, could create cascading effects, especially when immunologically unprepared species encounter novel or more virulent parasites. Altizer and her colleagues further pointed to a pattern of poleward and elevational range expansions of pathogens—an invisible frontier of invasion that puts already-stressed species at heightened risk. In some host-parasite systems, warming promotes faster development of parasite larvae and enhances vector survival, amplifying infection loads. Climate change is not just a driver of extinction but a "force multiplier" for parasites. While their study focused on modern biodiversity, the implications ripple backward: In times of ancient climate upheaval, the same mechanisms may have played out. Parasites, tracking heat, thriving in new habitats, and exploiting immunologically compromised species, may have acted as silent saboteurs in extinction events, nudging vulnerable populations past the brink.

Imagine a warming tundra enabling a tick species to invade the range of a declining herbivore. The tick carries a blood parasite that historically never encountered this host. This herbivore, already weakened by reduced food availability, begins dying off from anemia and sepsis. A local decline becomes regional, the regional becomes global, and soon a lineage vanishes—not solely from ticks, but tipped over the edge by them. Now imagine a different scenario: a warming northeastern United States, where rising temperatures allow malaria-carrying mosquitoes to expand their range into New England. Malaria was once endemic here, with outbreaks occurring as far north as Boston in the nineteenth century.[6] With warmer winters and wetter summers, *Anopheles* mosquitoes might once again find suitable breeding grounds. In a future shaped by climate instability and human migration, even a few introductions of *Plasmodium* protozoa from travelers or climate refugees could reseed the parasite. As immune-naive populations encounter a disease thought long gone, we may be forced to reckon with the return of ancient afflictions in new territories.

Parasites, by their very nature, are elusive in the fossil record. They lack bones, shells, or other durable tissues that typically fossilize well. But in rare and remarkable cases, they leave behind traces: a cyst entombed in mummified flesh, a cluster of eggs preserved in petrified dung, or, in one case, the wriggling ghost of a tapeworm larvae fossilized inside the intestinal tract of a Permian-era shark. This 270-million-year-old find, one of the oldest examples of an endoparasite, offers a visceral glimpse into just how long parasites have stalked the digestive systems of vertebrates.[7] Before dinosaurs thundered across the Triassic, parasites were already embedded in the guts of Earth's earliest predators. And it is not a stretch to imagine even the most fearsome ocean giants being vulnerable. Picture

Megalodon, the apex predator of the Miocene seas, slicing through marine food webs with five rows of serrated teeth. Its strength and size were unmatched, but was it ever truly alone? No direct fossil evidence has been found, but it is biologically plausible that *Megalodon* harbored its own suite of parasites: tapeworms coiled in its massive intestines, copepods clinging to its gills, or flukes tucked behind its eyes. Even titans are ecosystems. If *Megalodon* had parasites, they may have followed its carcass into extinction, leaving behind no trace but the evolutionary silence that comes when a host vanishes.

More often than not, we detect ancient parasites not by their bodies but by the genomic wake they leave behind. In mammalian genomes especially, genes associated with immune function—like the major histocompatibility complex (MHC) discussed earlier in the book—toll-like receptors, and interferon pathways bear the hallmarks of strong positive selection during the Late Pleistocene. These are not random mutations but scars of evolutionary warfare. The pathogens of prehistory, though unseen, were shaping genomes in ways we are only beginning to decode. Take the case of the human TRIM5α gene (short for "tripartite motif containing 5 alpha"), which restricts retroviruses. Its rapid evolution in primates suggests that ancient viral epidemics drove selection long before the human immunodeficiency virus emerged.[8] Elephants also carry a surprising genetic arsenal: whole suites of immune genes duplicated many times over.[9] These expanded defenses—from antiviral TRIM proteins to enhanced pathogen-sensing receptors—may have helped them weather ancient disease pressures that proved devastating to many of their megafaunal cousins. Even Neanderthal and Denisovan genomes, extracted from fossil bone, reveal ancient introgressions of immune variants that likely helped modern humans adapt to foreign parasites after migrating out of Africa.[10]

These molecular traces raise an uncomfortable but important question: In the great evolutionary lottery of survival, was it strength or size that mattered most, or adaptability at the microscopic scale? Did species that survived extinction events do so not because they were the biggest or the fastest, but because they were immunologically nimble, better at detecting pathogens, tolerating infection, or surviving parasitic siege? In this light, parasites cease to be mere hitchhikers; they are evolution's quiet coauthors.

And yet, parasites remain largely invisible in the grand narrative of extinction. A missing species in the fossil record is often attributed to climatic shifts, predation, or sudden catastrophes. But what if some of these disappearances were less dramatic, drawn out over generations by unseen infections that slowly chipped away at fertility, juvenile survival, or the energy needed to migrate, hunt, or defend? It is a haunting thought. The end of a lineage may have been written not in volcanic ash or impact craters, but in fevers, lesions, and wasting bodies. Even today, disease does not just threaten life; it rewrites its trajectory. For example, amphibians across the globe are being silenced by chytrid fungus, a single-celled parasite responsible for the most devastating wildlife disease outbreak in recorded history.[11] The oceans are not spared either. In recent years, sea star wasting disease has ripped through the Pacific Coast, melting entire populations of sea stars into pale, collapsing tissue within days.[12] Once abundant species like the sunflower sea star have vanished from vast stretches of coastline. The culprit appears to be a viral or multifactorial pathogen, an unseen microscopic force capable of unraveling whole marine ecosystems. In the tidepools and kelp forests alike, the absence of sea stars has triggered cascading shifts—prey species exploding, algae choking out habitat. These modern events are not anomalies—they are echoes. They remind us that extinction is a story not only of external pressures, but of internal vulnerabilities exposed.

If these microscopic killers can cause such rapid devastation now, in an age of antibiotics and conservation strategies, what might they have done in the deep past, when species had no such defenses? We may never find the parasite that helped bring down the mammoths, or the tapeworm that drained the vitality of a great predator on the edge, but we are beginning to understand that many of these animals were not alone in their final moments. They carried with them ancient passengers, some newly acquired, others as old as the lineage itself, whose presence may have tipped the balance between survival and oblivion. In that way, parasites are not footnotes in extinction; they are the agents behind the curtain, directing collapse from within. As our tools improve, through paleogenomics, metagenomic sequencing of coprolites, and molecular analysis of mummified tissue, we are beginning to exhume these invisible chapters. The microbial histories of long-dead species are slowly becoming readable. And when those stories come into focus, the arc of extinction may look very different—a saga not of brute force and sudden annihilation, but of erosion from within: a quiet unraveling of immune defenses, a cascade of physiological failures, a lineage quietly closing its final chapter, not with a roar, but with a cellular sigh.

COEXTINCTION: WHEN HOSTS DIE, SO DO THEIR SHADOWS

Coextinction is the quiet arithmetic of loss. When a species disappears, the organisms that depend on it, such as parasites, mutualists, and commensals, can vanish too. In theory, this seems obvious; in practice, it is astonishing how rarely we count those disappearances. As has been repeated numerous times throughout this book, parasites make up an enormous slice of life on Earth, likely on the order of dozens of percent of described species, and perhaps more when you tally the lineages we have barely begun to catalogue. (A team of senior undergraduate students working in my lab in 2024 discovered three distinct lineages of a trematode previously thought to be a single species that infects the edible mussel in New England based on just a few months of genetic research.) Yet parasites seldom appear in recovery plans, biodiversity targets, or red lists. When we save a host and purge its parasites "for health," we may be erasing entire evolutionary histories as collateral damage.

Ecologists have been warning for years that coextinction may be among the most common forms of extinction, even if it remains difficult to document cleanly. The logic is unforgiving: The more host-specific a parasite is, the more its fate is tethered to that host's trajectory; the smaller and more fragmented the host population becomes, the more likely its affiliates are to wink out. Modeling and review work over the past decade have tried to formalize all of this, arguing that the "extinction debt" we owe parasites, species fated to disappear as their hosts dwindle, has been systematically underestimated.

The California condor offers a cautionary parable. In the 1980s, biologists pulled the last wild birds into captivity, a heroic intervention that ultimately brought the condor back from the brink. But the birds were also deloused, and with that routine veterinary step, we likely erased an entire species: the condor's host-specific chewing louse, *Colpocephalum californici*. This "conservation-induced extinction" is now a textbook example of collateral loss in species rescue. Regardless of the extent to which the louse harmed its host, it represented a lineage with its own evolutionary story, one that ended, rather unceremoniously, on a zoo exam table. In 1990, a brief letter appeared in the correspondence section of *Nature*, written by Donald A. Windsor of Norwich, New York:

> SIR—In an attempt to save certain species from extinction, for example the California condor, the black-footed ferret and so on, how much attention is being given to their natural parasites? When all of the last remaining members of a species are taken into captivity, they may lose their parasites, either by the drastic change in living conditions, by

> treatment from zoo veterinarians or by generations of captive breeding. When they, or their offspring are released back into the wild, will they be able to become reinfected? Some parasites are quite host-specific in the wild and may indeed become extinct when their natural hosts are gone. "So what?" may be a typical reaction.
>
> But, if our goal is to conserve biological diversity, then indeed all species should be considered, not just those with the most outward appeal. Many hosts evolved or better still co-evolved with their parasitic burden. Perhaps they deserve each other.
>
> Equal rights for parasites![13]

Read today, Windsor's words sound like a preamble to William Campbell's Nobel Prize speech in 2015. More than thirty years ago, Windsor was already pressing against the dominant paradigm, arguing that parasites were not disposable passengers but coauthors in the evolutionary narrative. His quip, "equal rights for parasites," still stings with irony, but beneath it was a radical suggestion: If biodiversity is what we aim to conserve, then parasites must count too.

Not every supposed coextinction holds up under scrutiny. The "passenger pigeon louse," *Columbicola extinctus*, was long assumed to have vanished with its famous host. Later work, however, revealed that the louse was alive and thriving on the passenger pigeon's close relative, the band-tailed pigeon.[14] It was an important reminder that parasites are not always as host-strict as we imagine, and that our stories of dependence can be muddied by incomplete natural history. We have already seen how this kind of blind spot complicated the global eradication campaign against the Guinea worm. At first, it seemed like a perfect demonstration of how severing a parasite from its sole host could seal its fate. Yet the discovery of an unexpected reservoir in dogs overturned that assumption, showing how even a narrow host range can be more elastic than it appears. The lesson cuts both ways: A few parasites may slip the noose by shifting hosts, but for many others, there will be no escape.

The case for taking parasites seriously is not just about tallying losses; it is ecological. Parasites modulate food webs, shape host abundance, and can even respond positively to ecosystem protection. Studies of marine reserves, for instance, have shown that safeguarding hosts can also "rewild" their parasite communities.[15] In other words, if you want functioning ecosystems, you inherit their parasites, too. That insight has catalyzed a movement: a "global parasite conservation plan" with concrete goals for data mobilization, risk assessment, and on-the-ground practice, as well as an emerging open database to track extinction risk in neglected symbionts. It is an attempt to move parasites from pest lists to priority lists.

One creative effort to operationalize this vision has been the first-ever "parasite bioblitz." Borrowing from the citizen-science tradition of biodiversity bioblitzes, in which volunteers catalog every visible organism in a defined area over a short period of time, researchers adapted the concept specifically for parasites. In April 2023, a team of parasitologists converged on Stono Preserve in coastal South Carolina for a twelve-day survey across wetlands, ponds, brackish impoundments, and tidal creeks. They screened hosts directly, dissecting fishes and invertebrates, and paired this with cutting-edge environmental DNA (eDNA) barcoding from water and sediment samples. The results were staggering: nearly a hundred species identified by traditional means and more than a thousand distinct genetic lineages detected through eDNA, most of them previously unknown to science.[16] Many of these were tightly linked to specific hosts, reminders of how vulnerable parasite lineages are when those hosts decline. More than just a catalog, the ParasiteBlitz showed that parasites could be studied with the same intensity and creativity usually reserved for birds, plants, or insects, and that they deserve a place in the front lines of biodiversity conservation.

This leads to an obvious question: How large is the shadow we are ignoring? Estimates vary, but conservative tallies suggest that well over a hundred thousand helminth species infect vertebrates alone, even before we count the staggering diversity of parasitic arthropods, fungi, protozoa, bacteria, and viruses.[17] Climate change projections add another layer of urgency: Using global occurrence records for hundreds of parasite species, one analysis forecast that 5 to 10 percent could face climate-driven extinction by 2070, with even larger fractions at risk once host range contractions and coextinction cascades are included.[18] These are both numbers and genealogies poised to evaporate.

Coextinction is already baked into how we manage endangered hosts. Captive breeding, translocations, reintroduction programs, and veterinary prophylaxis almost always include deworming and delousing. The intention is compassionate and often necessary for near-term survival, but over the long term, it can produce "ghost hosts": species restored without their coevolved symbionts, ecological copies stripped of a portion of their biodiversity. Some ecologists have even suggested reintroducing host-specific parasites after host recovery, a proposal that sounds provocative until you realize it is just consistency: If the goal is to restore ecological communities, parasites are part of the community.

What would it mean to take that seriously? Imagine a condor recovery plan that, alongside lead abatement and nest monitoring, included a strategy for maintaining its endemic lice in a controlled way; or an amphibian conservation project that, while fighting chytrid fungus, also tracks the fates of benign helminths that may play roles in immunity. This is the frontier the parasite conservation movement is carving out—not a naive "save every parasite" stance, but a targeted, evidence-based approach to prevent inadvertent erasure of lineages that regulate systems, store evolutionary information, and sometimes even buffer hosts against worse invaders.

The science is not at all settled, and humility is essential. Some parasite-host ties are looser than we think; others are tighter and more fragile. Parasites can be villains—drivers of disease outbreaks and contributors to collapse. But they are also archivists of evolution, keeping meticulous records in their genomes and life cycles of who lived with whom, and for how long. Cutting them out of conservation narratives reduces biodiversity to large, photogenic vertebrates and a handful of charismatic pollinators. It also blinds us to the quiet scaffolding that holds ecosystems together. And it invites a larger philosophical question: What do we mean by biodiversity? If we mean only the species we find pleasing or useful, then coextinction is a footnote. If we mean the totality of life's experiments, the winners and their hangers-on, the intimate dependencies that knitted complexity from conflict, then parasites must be counted. Saving hosts while letting their affiliates disappear is not a victory; it is a partial reconstruction, a museum diorama where the background has been painted over.

UNEARTHING THE UNSEEN: THE RISE OF PALEOPARASITOLOGY

Picture the opening sequence of a Spielberg film. Instead of a mosquito trapped in amber yielding dinosaur DNA, a scientist extracts sediment from a thirty-thousand-year-old lake bed. Under the microscope, parasite eggs emerge—silent remnants of infections that haunted an earlier ice age. The premise is no longer dinosaurs crashing through the jungle, but pathogens resurrected from the deep freeze of time: a whipworm crawling from a coprolite, a protozoan flickering to life in a thawed mummy's gut. What would such a screenplay look like? Perhaps it would be less spectacle and more horror: parasites slipping quietly into modern hosts, immune systems unprepared for foes last seen in the Ice Age. Unlike the lumbering giants of Spielberg's imagination, these resurrections would not roar; they would whisper, spread, and infiltrate. In the age of pandemics, such a film might feel more documentary than fiction.

That eerie possibility is precisely what gives paleoparasitology its strange power. It is the science of resurrected ghosts—not in the sense of reanimating them, but in recovering their stories. It is, by necessity, a science of scraps. Parasites rarely fossilize whole. Most are soft-bodied, fragile, and hidden inside other organisms. Yet sometimes they leave behind traces so stubborn that time cannot erase them. Coprolites—fossilized feces—have offered some of the most evocative glimpses. Under the microscope, a whipworm egg from the Pleistocene or a hookworm larva from a seven-thousand-year-old site in South America appears almost identical to its modern descendants, its shell still intact, its outline unmistakable. These minuscule ovals, no larger than a grain of sand, have survived where entire species have not. A coprolite is both proof of infection and a time capsule of diet, hygiene, and movement. In Viking latrines in Denmark, whipworm and giant roundworm eggs revealed that intestinal parasites flourished in medieval northern settlements despite the cold climate.[19] In the dry caves of the Andes, two-thousand-year-old mummies yielded hookworm and whipworm eggs, proof that these tropical parasites were already entrenched in the Americas long before Europeans arrived.[20] Such finds challenge our assumptions about migration and contact, showing that parasites followed, and sometimes revealed, the hidden routes of human history.

Other clues are more dramatic. A tick encased in amber, still gripping reptilian skin from a Cretaceous forest, speaks to the deep antiquity of blood feeding. Some specimens even hold traces of hemoglobin, the last meal sealed inside, as if time had conspired to preserve an evolutionary crime scene. Such fossils are astonishing, but they are also reminders of how skewed the record is. Parasites rarely fossilize whole; only their hardiest stages, their eggs and cysts, persist, leaving the history of delicate protozoans like malaria, leishmania, or trypanosomes largely invisible. For decades, this bias misled scientists into thinking these lineages must be more recent than they are. The archaeological record helped fill the gap. Medieval burials in Europe harbor whipworm and roundworm eggs. In pre-Columbian South America, mummies bear evidence of Chagas disease, caused by *Trypanosoma cruzi*.[21] These discoveries are both curiosities and cultural artifacts. They illuminate how human settlement, agriculture, and trade amplified the reach of parasites, how invisible burdens shadowed empires and reshaped societies.

Then came the molecular revolution. Ancient DNA transformed paleoparasitology from a field of fragments into one of whole narratives. But the shift was anything but easy. DNA is a fragile molecule, prone to breaking apart in the presence of heat, oxygen, water, and time. What paleoparasitologists usually recover are not intact genomes but shattered relics: short fragments, often chemically modified, riddled with errors, and mixed with vast quantities of microbial and environmental DNA. To coax meaning from these scraps requires an arsenal of technologies developed in the past two decades. Polymerase chain reaction (PCR) was the first tool to make a dent, allowing scientists in the 1990s to amplify tiny fragments of DNA from mummies and coprolites. But PCR is a blunt instrument; it can only target specific, already suspected sequences, and it is prone to contamination. For years, claims of ancient DNA findings were met with skepticism. The real breakthrough came with next-generation sequencing (NGS), which allowed entire libraries of fragments to be sequenced in parallel. Instead of needing long, pristine stretches, researchers could now stitch together overlapping snippets, like piecing a shredded manuscript back into legible form.

Equally transformative were hybridization capture techniques, in which scientists design "bait" molecules that bind only to DNA fragments matching a parasite's genome. This method acts like a magnet in a haystack, pulling parasite DNA out of the overwhelming background of bacterial, fungal, and host DNA. In the case of a high-Himalayan discovery of *Plasmodium falciparum*, such targeted capture was the only way researchers could detect malaria's faint genetic fingerprints among billions of irrelevant sequences. Even more recently, bioinformatics advances have rewritten the field. Computational pipelines can now distinguish authentic ancient DNA, often characterized by chemical scars like cytosine deamination, from modern contaminants. Artificial intelligence algorithms designed to detect these unique damage patterns ensure that what we call "ancient malaria" or "ancient Giardia" is not simply a lab worker's sneeze contaminating a sample.

The result is nothing short of revolutionary. In the Americas, coprolites that once yielded only visible helminth eggs are now mined for hidden protozoans like *Giardia*, organisms too delicate to fossilize. Ancient sediments from lake beds, long assumed sterile, now give up parasite DNA thanks to shotgun sequencing and metagenomics, revealing the rise and fall of parasite communities as climates shifted thousands of years ago. What is striking is not just that these tools work, but how improbable their success seemed only a generation ago. Paleoparasitology is built on the most fragile evidence imaginable: broken strands of DNA that have persisted in mud, bone, or feces for millennia. And yet, from these fragments, entire evolutionary histories are being rewritten.

The implications are profound. Paleoparasitology no longer just proves that parasites were present; it can reconstruct entire communities. Ancient feces can now be read like ecological archives, showing which species shared a gut, how diverse parasite loads were, and how they shifted through time. This capacity opens entirely new windows: We can ask not only when parasites emerged, but how they coevolved, how they vanished, and how their fates intertwined with those of their hosts.

What began as a fringe pursuit, counting worm eggs in fossilized droppings, has matured into a discipline that links archaeology, genomics, ecology, and conservation. It reshapes how we understand ancient disease burdens, human migrations, and even extinction dynamics. By studying how parasites shifted ranges during past climate upheavals, scientists can better anticipate how they will respond to global warming today. By tracing parasite extinctions in tandem with their hosts', we learn how tightly bound biodiversity really is, and what is lost when species vanish. Above all, paleoparasitology demands respect for the smallest actors in evolution's drama. They rarely fossilize, but their absence distorts the record as much as their presence enriches it. With every egg in a coprolite, every genome coaxed from a mummy, every lesion in an ancient bone, the field insists on a truth too long overlooked: that parasites were never just passengers. They were drivers, coauthors, sometimes executioners. And their ghosts, recovered from the scraps of history, may yet change how we tell the story of life on Earth.

INTERVIEW FEATURE: MEGAN MICHEL

Science is not just about tools; it is about the people who wield them. If paleoparasitology has matured from a fringe pursuit into a discipline capable of rewriting evolutionary history, it is because a new generation of researchers has learned to coax entire narratives out of the faintest molecular traces. Few illustrate this more vividly than Megan Michel, a doctoral candidate jointly affiliated with the Department of Human Evolutionary Biology at Harvard University and the Archaeogenetics Department at the Max Planck Institute for Evolutionary Anthropology, whose work has pulled malaria out of the Himalayas' icy silence and into the scientific record. Her work struck me personally and professionally. More than a decade earlier, as a graduate student in South Africa, I had found myself in a pub on a rainy June night, debating with colleagues about whether the golden age of discovery had passed. Some argued that breakthroughs like the double helix or the human genome project belonged to another century. I countered that science had not slowed; it had fractured into dozens of new frontiers. Paleoparasitology was one of them, though it was barely on anyone's radar then. A decade later, listening to Megan describe how she and her collaborators pieced together the faint genetic whispers of *Plasmodium falciparum* from ancient Himalayan remains, I realized how prescient that argument had been. What once seemed like scraps—too fragmented, too degraded, too improbable—are now the building blocks of discoveries as profound as any in biology.

By the time I reached out to Megan, the field of paleoparasitology had already shifted from scraps of eggs in coprolites to full-blown genomes reconstructed from the most improbable remains. But her work, published in *Nature* in 2024, pushed that transformation further than anyone expected. The study, which spanned 5,500 years of human history and sixteen countries, successfully recovered genome-wide data from *Plasmodium falciparum*, *P. vivax*, and *P. malariae*, rewriting our understanding of malaria's antiquity and its global spread.[22]

When I asked Megan how she found herself here, her answer was disarmingly modest. She never set out to be a parasite detective. As an undergraduate, she completed her major in biology and minor in history, chasing the intersection of two worlds: the hard science of life and the stories that frame it. A job as a technician in an ancient DNA lab launched her trajectory, and soon she was grinding skeletal remains into powder, coaxing out fragments of genomes. The more she handled those fragments, the more she realized what was missing from the story of human evolution: the parasites. "I've always been drawn to the selective pressure malaria has had on the human genome," she told me. "It's one of the strongest forces we can see, written into our DNA."

Her most surprising result came from a skeleton excavated at Chokhopani, a high-altitude site in Nepal more than 2,800 meters above sea level—well above the range of anopheline mosquitoes. At first, she assumed it was a false positive. But targeted enrichment via hybridization capture revealed authentic *P. falciparum* DNA. The implications were stunning. This three-thousand-year-old individual, nicknamed CHO001, carried malaria not because it circulated locally, but because people were moving. The infection was likely contracted in lower, mosquito-ridden valleys and carried back into the mountains, effectively turning malaria into a tracer of trade, travel, and cultural exchange.

"It really forced us to think differently," Megan explained. "Mobility wasn't just about goods—it was about pathogens. Diseases traveled with people and reshaped the genetic landscape in ways we're only now beginning to untangle." Her point struck me: Paleoparasitology is not just about parasites; it is about human connectivity. Every infected tooth or bone fragment becomes a map of ancient roads, markets, and migrations.

The *Nature* study also provided the first high-coverage ancient genomes of malaria species, showing *P. vivax* circulating across Europe by the fourth millennium BCE—thousands of years before the earliest written references. These findings not only predate Hippocratic accounts of fevers in Greece but reveal malaria's deep entrenchment in prehistoric Europe. They also clarify how malaria arrived in the Americas: *P. vivax* strains closely matched those from Europe, introduced during colonization, while *P. falciparum* clustered with African lineages brought across the Atlantic through the slave trade. When I asked Megan whether parasites shaped us as much as climate or food, she did not hesitate. "It's almost impossible to separate those factors. Scarcity, stress, violence, parasites—they worked together."

But it appears that malaria in particular has left one of the clearest fingerprints on our genome. Sickle-cell disease, G6PD deficiency, thalassemia—those are not random quirks. They are scars of survival. For her, ancient DNA is not just a way of filling gaps; it is the only way to access populations and parasite lineages that no longer exist. Modern strains tell us what survives; ancient DNA resurrects what has been lost. It shows us how malaria was present in Bronze Age Europe, how it reached Himalayan valleys, how it helped redraw the genetic contours of humanity.

As our conversation wound down, Megan mentioned her excitement at presenting at the International Symposium on Biomolecular Archaeology in Turin, Italy. For her, it was another milestone in a field that, not long ago, barely existed. For me, it was a reminder of how fast science splinters into new frontiers. Disciplines that were not even in the vocabulary of my graduate training are now standing on the main stage of international conferences. Paleoparasitology, archaeogenetics, biomolecular archaeology—each represents a widening of the lens through which we view the past. What struck me most was not just Megan's technical brilliance, but the quiet conviction that parasites, too, belong in humanity's story. Her work is proof that even the faintest genetic whispers from a three-thousand-year-old body can alter how we think about disease, migration, and survival. In a time when our own future feels precarious, these voices from the deep past remind us that the smallest actors often shape the grandest narratives.

GENOMES AS TIME MACHINES: WHAT ANCIENT DNA REVEALS ABOUT PARASITE EVOLUTION

Listening to Megan describe how a three-thousand-year-old skeleton from the Himalayas could still whisper the story of malaria made something clear: DNA is both a tool for archaeology and a time machine. Each fragment of a sequence is a postcard from a vanished world. When enough of these postcards are pieced together, they allow us to redraw the evolutionary maps of parasites that have shaped life for millions of years.

For decades, parasite phylogenies were built almost entirely on modern lineages. We could compare *Plasmodium* strains in Africa, Asia, and South America; we could track *Leishmania* diversity across continents; we could catalog the genetic signatures of helminth parasites in humans and livestock. But these reconstructions were always incomplete. They lacked the ghosts—the lineages that once existed, then blinked out, leaving only gaps in the tree. Ancient DNA fills those gaps. It extends the branches backward in time and reveals branches and twigs we never knew were there.

Take malaria. By recovering *Plasmodium vivax* from Bronze Age Europe and *P. falciparum* from Himalayan highlands, researchers have shown that malaria was not a recent scourge of the tropics but a global traveler far earlier than written records suggest. Ancient DNA lets us glimpse extinct strains, revealing that the parasites of the past were not carbon copies of those we see today. They were genetically distinct populations, some lost forever, others absorbed into the lineages that still circulate. These discoveries change how we think about the evolutionary arms race between parasites and hosts. The sickle cell allele in humans, for instance, is not simply a response to malaria in the abstract, but to specific strains that rose, spread, and disappeared over time. Without ancient genomes, that context would be invisible.

The same logic applies to other parasites. Ancient DNA from *Trypanosoma cruzi* in mummified South American remains pushes the antiquity of Chagas disease back thousands of years, linking it to early agricultural and settlement patterns. *Leishmania* DNA recovered from renaissance Italy places visceral leishmaniasis squarely in European history, countering assumptions that it was a purely tropical disease.[23] Even helminths, which leave behind eggs, now yield genetic material that can be compared across centuries, revealing not only persistence but also turnover—extinctions and replacements invisible to the naked eye.

But context is everything. A parasite genome alone tells only part of the story; its true meaning emerges when paired with archaeology, ecology, and climate records. Was a mummy infected with hookworm because agriculture created stagnant irrigation channels that allowed larvae to thrive? Did a shift in animal domestication expose humans to novel tapeworm strains? Was a sudden spike in malaria DNA in lake sediments tied to a warming climate that expanded mosquito breeding grounds? Only by integrating these threads can paleoparasitology live up to its potential—not simply cataloguing the past, but explaining it.

The challenges remain immense. Intestinal worms, for instance, often degrade beyond recognition unless preserved in the driest or coldest conditions. Even the chemical stability of DNA cannot withstand the full onslaught of time. Protozoa, with their fragile cell membranes, are almost impossible to recover unless captured in coprolites or specialized sediments. Preservation biases skew what we can see, and much of parasite history is still invisible. Yet every technical advance, from shotgun sequencing to targeted capture, from improved damage authentication to machine learning algorithms, pushes the window a little further open.

The future promises even more. Larger sample sizes will allow us to reconstruct parasite population structures through time, not just isolated infections. Higher-resolution genomes may let us detect how parasites adapted to shifts in host immunity, diet, or migration. And as databases of both ancient and modern strains grow, we may begin to track not only parasite extinction but parasite evolution in real time, across thousands of years. In that sense, paleoparasitology is both retrospective and predictive. By watching how parasites moved, diversified, or vanished in response to ancient climate upheavals, we gain clues about how they might respond to the unprecedented instability of the Anthropocene. The past is not only a ghost story; it is also a guidebook.

GHOSTS THAT STILL HAUNT US: ANCIENT PARASITES AND MODERN HEALTH

Imagine this: a toddler in a gleaming modern hospital, hooked up to machines, fighting a sudden storm of asthma. Her lungs are inflamed, her immune system overreacting to something as trivial as pollen. To her doctors, it is an everyday case. To an evolutionary biologist, it is a ghost story. The child's wheezing might be traced, in part, to parasites she has never met. For most of human history, intestinal worms like *Ascaris* and *Trichuris* were constant companions. They trained our immune system, teaching it when to respond and when to stand down. In their absence, the immune system sometimes panics, treating a grain of pollen as if it were a legion of worms. Her asthma attack is a faint echo of evolutionary battles long lost, an immune system haunted by partners that have gone missing.

This is the unsettling truth of infectious disease research: The wars of yesterday are still shaping our health today. Our DNA carries the memory of ancient infections, engraved in the form of mutations, immune strategies, and trade-offs. They are evolutionary graffiti—reminders that survival is not just about predators and prey, but about invisible battles that rages on at the cellular level. Take the Duffy antigen discussed in the previous chapter. To recap, this antigen is a protein found on the surface of red blood cells. In much of sub-Saharan Africa, a single genetic variant silences the

Duffy receptor. For millions, this genetic absence means safety: *Plasmodium vivax*, one of the main malaria parasite species, cannot get inside their red blood cells. But this shield comes with consequences. The receptor is involved in other immune pathways, and without it inflammatory responses can take strange turns. What looks like a win against malaria may quietly increase vulnerability elsewhere. That is how evolution works—not with clean victories, but with compromises carved into our genomes.

Sometimes those compromises come back to haunt us. During the COVID-19 pandemic, scientists found that people who carried a specific cluster of immune genes were far more likely to develop severe disease. The surprise came when geneticists traced that cluster back tens of thousands of years, to Neanderthals.[24] In Ice Age Eurasia, these immune variants may have been lifesaving against prehistoric pathogens; in 2020, they became a liability. A pandemic turned into an archaeological excavation of our DNA, unearthing the ghosts of ancient epidemics that still determine who lives and who dies.

Not all of these ghosts are molecular. Some appear in unexpected corners of modern medicine. Chronic infection with schistosomes, the blood flukes that burrow into human veins, has plagued communities for millennia. Archaeologists have found their eggs in six-thousand-year-old remains in the Middle East, proof that they have shadowed us since the dawn of cities.[25] Today, the infection is not just debilitating in the short term; it also leaves a carcinogenic footprint. In regions where schistosomiasis is common, bladder cancer rates are strikingly higher.[26] Over centuries, a parasite that once merely drained energy sculpted patterns of cancer risk.

Even our public health tools are haunted. Vaccines that work perfectly well in Europe often falter in parts of Africa or Asia.[27] At first glance, it looks like a failure of technology, but if we dig deeper, we see the explanation lies in parasite history. Populations that have lived for generations with heavy helminth burdens have immune systems tuned differently. Worm infections push the immune system toward tolerance, dialing down inflammation so that hosts do not self-destruct. That "training" lingers long after the worms are gone, dampening vaccine response. Once again, the ghosts of parasites tilt the scales of modern health. There are other stories in which the ghosts do not merely whisper; they scream. Consider the rise of helminthic therapy, born from the hygiene hypothesis, first proposed by the British epidemiologist David Strachan in 1989. This idea suggests that our sanitized, germ-free environment, so different from the one our immune system evolved in, has left our defenses overreacting. Parasites, especially worms, were once constant companions, tuning our immune system like a conductor guiding an orchestra. Without them, our immune responses can go haywire, mistaking harmless substances for threats and leading to allergies, asthma, and autoimmune disorders.

In 2010, doctors took the leap from theory to practice in a bold, controversial series of clinical trials: They deliberately infected patients with microscopic whipworm eggs (*Trichuris suis ova*) to treat Crohn's disease and other inflammatory bowel conditions.[28] The logic was audacious but simple: If our immune systems evolved alongside worms, perhaps reintroducing them could calm autoimmune turbulence. Some patients improved; others experienced no benefit. But for many, the treatment triggered weeks of gastrointestinal distress: diarrhea, cramps, gas. It was as if the body recognized the old visitor but did not know how to handle their arrival.

The story did not end there. On the fringes, self-treatment entered a strange and perilous territory. Some companies began selling pig whipworm eggs directly to consumers—no prescription, no medical supervision. One company even provided instructions for use online, branding itself as "all natural" (so obviously it must be good). The pitch? Restore your immune balance by swallowing a vial of worm eggs. This might sound like a bad 1980s horror flick, but it is real—and dangerous. A 2017 *Men's Health* article recounts how such products skirted regulatory oversight by marketing themselves as food additives in Germany.[29] Customers, acting without medical guidance, attempted this therapy on their own. Experts emphatically warned that this is not advisable: "Self-medication with any type of worm is not recommended . . . may cause quite severe side effects if not monitored carefully by a doctor."

Scientific guidelines reinforce that concern. Clinical studies on controlled helminth therapy show frequent side effects in the first months, with moderate to severe gastrointestinal symptoms compared with placebos.[30] These symptoms often lasted days to weeks. Worse still, immunity is not guaranteed, and long-term colonization of these worms within the human body remains unpredictable at best.

This is where the hygiene hypothesis meets hard reality. The idea is not wrong—parasites do modulate our immunity—but the solution cannot be simply swallowing worms (although this is not nearly as negligent as cases of people swallowing tapeworms for weight loss, resulting in severe illness and, in some cases, hospitalization).[31] The therapy itself risks malnutrition, anemia, infection, and unknown long-term consequences. Many scientists now argue that the future lies not in reviving parasitic infestations but isolating the beneficial molecules they secrete. These immunomodulatory compounds could offer the same immune-calming benefits without the need for self-parasitizing.

In other words, our immune systems may indeed be haunted by the absence of ancient parasites, but self-parasitizing is not the exorcism. It is a desperate fix loaded with risk. Like dangling a ghost from your ceiling in hopes it will do your housework, it may soothe the spirit, but it could also break a window—or worse.

PARASITES IN CONSERVATION AND THE POLITICS OF EXTINCTION

Parasites are rarely invited into the conversation when conservationists draw up recovery plans. They are cast as villains, irritants, or at best background actors in the larger drama of ecosystems. When a species teeters on the brink, our instinct is to strip it of its burdens: delouse it, deworm it, medicate it clean. On one level, this seems compassionate, even necessary. But at the species level, these interventions can carry a hidden cost: the quiet erasure of parasites that have evolved alongside their hosts for millions of years. As we saw earlier, the California condor's chewing louse may already be gone, snuffed out not by extinction of its giant host but by the well-meaning hand of conservationists. It is an unsettling irony: In saving one species, we may have condemned another, uncounted and unmourned. This kind of collateral loss, what some have called conservation-induced extinction, forces us to reconsider what biodiversity really means. For decades, parasites were treated as noise in ecological systems, irritants complicating the elegant balance of predator and prey. But the science has changed. Food webs that include parasites reveal far more connections than those that ignore them. In estuaries, the biomass of trematode larvae released from snails has been found to rival that of top predators.[32] Parasites redistribute energy, regulate host abundance, and link species in subtle, powerful ways. To erase them is to pluck out threads from a web we barely understand.

Yet the ethics remain uneasy. Should we save parasites when they harm the hosts we are fighting to protect? The question divides conservationists. On one hand, parasites are archivists of evolutionary history, their genomes chronicling past battles and migrations. On the other, they can be lethal, driving populations already endangered by human activity further toward collapse. The contradiction is so stark that legal scholars now describe parasites as an "uncomfortable category" in conservation law: organisms simultaneously life-threatening to humans and livestock, yet among the most diverse and ecologically significant on Earth.[33] The Convention on Biological Diversity technically protects them, but in practice they are absent from endangered species lists, recovery targets, and funding priorities. Our laws mirror our biases: Large, photogenic and charismatic animals are counted; parasites are erased.

A few scientists are trying to change that. In 2020, a coalition of parasitologists proposed a Global Parasite Conservation Plan, arguing that parasites are "among the most diverse, threatened and underprotected animals on Earth."[34] Their plan set goals for data mobilization, risk assessment, and global monitoring networks. The first parasite bioblitz in 2023, described earlier in this chapter, uncovered a treasure trove of undiscovered biodiversity within a relatively confined area. But the bioblitz also sharpened the question: What do we lose if we let parasites go? The easy answer is "nothing." Few would mourn the disappearance of malaria or the Guinea worm. Yet such a view ignores the broader role parasites play. Many infections are chronic but nonlethal, shaping immune systems, regulating populations, and even buffering hosts against more virulent invaders. Their genomes preserve the deep history of host-parasite relationships, history we cannot recover once they vanish. To dismiss their extinction as a "good riddance" is to risk repeating the mistake we once made with predators—believing their loss would simplify ecosystems, only to discover it destabilized them.

The politics of parasites is, at its core, a politics of values. What kind of biodiversity are we trying to save? If conservation is only about the charismatic and the useful, parasites will always be expendable. But if it is about preserving the full tapestry of life, then parasites must count too. Some ecologists have gone so far as to suggest reintroducing benign, host-specific parasites to recovered species—a provocative idea, but one rooted in ecological consistency.[35] If we mean to restore ecosystems, should we not restore them whole? This raises an even stranger, almost science-fictional question: If we are willing to talk about de-extinction for mammoths, dodos, passenger pigeons, and more recently dire wolves, why not their parasites? When the California condor was pulled into captivity in the 1980s and its louse disappeared in the process, we lost more than an insect. We lost an evolutionary lineage that had coexisted with condors for millennia, adapting in parallel to their feathers, their roosting habits, their very biology. The condors that circle the skies today are genetically the same species, but in some subtle sense, they are ecological amputees—returned without their shadow fauna.

Technically, resurrecting a parasite might be easier than resurrecting a mammoth. A louse genome is small. Museum skins, amber specimens, or old captive birds may actually contain enough DNA fragments to reconstruct a genome. With next-generation genetic engineering and synthetic biology, one could, in theory, engineer a close relative into a proxy host and bring the extinct louse back. Tools already exist in principle; what is missing is not science but will.

Here is where the ethics get thorny. Would anyone fund a program to bring back a louse? Could you imagine the headlines? We already know from movies like *The Thaw* how the general public feels about microbes and parasites reemerging from the past. To most conservationists, the very suggestion sounds absurd. Why resurrect a parasite whose entire ecological role seemed to be irritating its host? Yet if the goal of de-extinction is authenticity, restoring ecosystems as they truly were, then ignoring parasites undermines the project. A condor without lice is not the condor that evolved for millennia. It is a sanitized copy, a creature curated for human aesthetics. The ecological case for parasites, even the tiny "nuisance" ones, is stronger than most people realize. A host stripped of its natural parasites might be more vulnerable to novel pathogens, much as an immune system deprived of training becomes hypersensitive to allergies. If we are willing to clone trees to restore forests, or rewild wolves to stabilize prey dynamics, why stop at parasites? Their absence is as much a distortion of ecosystems as the loss of a predator.

But the risks are real, too. What if a resurrected parasite is less host-specific than assumed? What if condors, weakened by decades of bottlenecks and captivity, cannot tolerate their old affliction? A reintroduced parasite could, in the worst case, tip a fragile species back toward collapse. The argument quickly turns into a philosophical mirror: Is biodiversity a numbers game, in which every lineage is sacred, or is it a pragmatic project in which some lives matter more than others? De-extinction of parasites also forces us to interrogate the romance that drives so much of conservation. When we dream of woolly mammoths thundering across Siberia again, we do not imagine them coughing from lungworms or carrying lice through their shaggy coats. Yet that was part of their reality. Bringing back the mammoth without its parasites would be like restoring a cathedral without its gargoyles: cleaner, perhaps, but incomplete. If the purpose of de-extinction is to make the past real again, then parasites are part of that reality.

Is it worth it? For now, the question remains purely academic. No one is drawing up proposals to revive extinct lice. But the thought experiment is revealing. It exposes how selective our commitments to biodiversity really are. We say we want to save "life on Earth," but in practice we save what is beautiful, useful, or beloved. Parasites force us to confront the harder truth: If conservation is about wholeness, then we must be prepared to save, and perhaps even bring back, the lives we do not love. A de-extinct condor with no louse is a painting restored without its shadows—perfect on the surface, but missing the texture that made it real. Whether we are willing to fill in those shadows will say more about us than it does about the parasites.

Perhaps the hardest truth is that ignoring parasites is itself a choice, a selective storytelling about nature that flatters our aesthetics while misrepresenting reality. Future generations may look back on our blindness with the same bemusement we reserve for Victorians shooting birds for hat feathers. Or they may see it as a necessary compromise, sparing hosts even at the expense of their shadows. Either way, the decision matters. Parasites are not afterthoughts. They are not simply villains. They are coauthors of life's story, and their extinction is as real, and as consequential, as that of any host.

PUTTING IT ALL TOGETHER: SHADOWS IN THE FOSSIL RECORD

Extinction stories often end with silence: the last whale's song cut short, the final tree felled, the last bear disappearing into the void of numbers. But what we so often forget is that when one species vanishes, it rarely goes alone. It drags entire worlds down with it: the mites in its fur, the worms in its gut, the viruses in its blood. These are the shadow lineages, lives bound so tightly to their hosts that their existence is eclipsed until it is already too late. To speak of extinction without them is to tell only half the story.

Parasites are the memory keepers of evolution, the ghostwriters in nature's margins. They do not just accompany species; they inscribe relationships—who fed on whom, who traveled where, who adapted in response. Every whipworm egg in a coprolite, every trematode scar in a fossilized liver, every fragment of DNA coaxed from a mummy is a reminder that parasites record history from the inside out. They are intimate witnesses to survival and collapse, custodians of stories that bones alone cannot tell.

For decades, those stories were fragmentary. Paleoparasitology could offer glimpses—a worm egg here, a lesion there. It was enough to prove presence but not to trace legacies. Then the field became exposed to the molecular revolution. Suddenly, the silence of millennia began to crack. Shattered strands of ancient DNA, stitched together with modern sequencing, revealed parasites in places no one expected: malaria in Bronze Age Europe, *Giardia* in pre-Columbian feces, tapeworms coiled in the guts of sharks that swam long before the first mammals. These finds rewrote timelines, showing that parasites did not merely haunt humanity in recent centuries but shaped us across deep time.

Sometimes, all it takes is a single fragment to change everything. When Megan Michel and her colleagues recovered faint traces of *Plasmodium* from a three-thousand-year-old high-altitude mummy in Nepal, the result seemed impossible. Mosquitoes do not survive at such elevations. Skeptics thought it must be contamination, a false positive. But the DNA held up under every test. The only conclusion left was that this host returned, or was carried back, to the heights of Chokopani, where they died. That one whisper of parasite DNA reframed how we understand mobility, trade, and disease in the ancient Himalayas. It told us that parasites move with people and, in turn, tell us where people moved. A parasite's genome can be a map inked in blood.

The lesson resonates far beyond archaeology. If parasites are this deeply intertwined with our past, their disappearance in the present reshapes futures in ways we may not yet grasp. When we lose them, we do not just lose irritants; we lose regulators of food webs, shapers of immune systems, hidden threads of connectivity. We erase parts of the evolutionary script, sometimes entire subplots, and the story we are left with is thinner, more fragile.

What does it mean to carry these ghosts forward? Perhaps it is to listen more carefully to what the scraps are telling us—that biodiversity is not just lions, bees, or coral reefs. It is the lice, the flukes, the worms, the viruses, the entire unseen ecosystem that comes bundled with every living thing. Conservationists, paleontologists, and geneticists alike are beginning to accept that truth, however reluctantly. Parasites do not fit neatly into our categories of "good" and "bad." They are evolutionary forces, often cruel, sometimes protective, but always instructive. Which brings us back to the haunting question: If we are only now learning to hear the whispers of parasitic history, what else have we silenced without ever knowing it? How many evolutionary companions vanished before science had the tools to notice? And what would it mean to restore not just species, but the full, messy communities of relationships that made them whole?

The next chapter looks forward, not back. If parasites have ghosted their way through our evolutionary past, what roles will they play in our future? Climate change, shifting habitats, and human medicine are rewriting the rules again. Some parasites may vanish before we can name them; others may surge in new and alarming ways. From eradication campaigns like Jimmy Carter's crusade against the Guinea worm to the unsettling potential of parasites as bioweapons in an age of genetic engineering and the increasing sophistication of artificial intelligence, the puppeteers of evolution are stepping onto a new stage. Whether we fear them, save them, or reinvent them, one truth remains: Life on Earth will never be parasite-free. The question is whether we are ready to face what their survival—or their silence—will mean.

9

THE FUTURE OF EVOLUTION'S PUPPETEERS

The crisis consists precisely in the fact that the old is dying and the new cannot be born; in this interregnum a great variety of morbid symptoms appear.

—Antonio Gramsci, *Prison Notebooks*

Over the course of this book, I have argued that parasites do more than sell themselves. They are the hidden architects of life. They rewrote genomes and sculpted immune systems; they shaped civilizations through disease and famine; they left their fingerprints on our myths and our metaphors; they even helped construct the basic architecture of cells and ecosystems. They are not marginal characters in the story of life but its most persistent authors.

We now find ourselves in a strange new chapter of that story. Some scientists call it the Anthropocene—an age in which human activity is the dominant force reshaping the planet. Forests fall, ice sheets melt, species migrate, genomes are edited, and artificial intelligence begins to design alongside us. Parasites will not be exempt from this upheaval. In fact, they may be its most sensitive witnesses and its most cunning opportunists. A warming world means new ranges for ticks and mosquitoes; a world of global trade means new corridors for invasive worms and viruses; a world of genetic engineering means parasites may become tools—or weapons—in human hands.

Parasites have always shaped events from the background as puppet masters, tugging at hidden strings of bodies, ecosystems, and history. The question now is what happens when humans attempt to seize those strings for themselves. Will we cut them, strengthen them, or tie them into knots we do not yet understand? The future of evolution's puppeteers is also the future of us, because in truth, there has never been a clear line between the puppeteer and the puppet.

THE ANTHROPOCENE AS A NEW ARENA

The Anthropocene has been called many things: the age of humans, the age of industry, the age of collapse. It is often described as an epoch of crisis for the natural world. Yet for parasites and pathogens, it is also a time of extraordinary opportunity. The forces reshaping the biosphere, such as climate change, urbanization, industrial agriculture, and globalization, are expanding the arenas in which parasitism plays out. Parasites are opportunists, and never before have humans created so many openings for their proliferation.

Climate change is already rewriting the distribution map of parasites. Rising temperatures and shifting seasons extend the reach of malaria-bearing mosquitoes into highlands that once served as refuges. Ticks that transmit Lyme disease are expanding northward into Canada and deeper into Europe as winters shorten.[1] Marine parasites, too, are moving. Protozoans like *Perkinsus marinus*, which causes massive mortalities in oyster populations, once confined to the warm southern United States, have moved northward into cooler estuaries, aided by warmer waters that no longer check their reproduction.[2] In my own lab, a team of undergraduate students recently discovered a unique genetic variant of the trematode *Proctoeces maculatus* infecting blue mussels in coastal Maine.[3] Prior to this research, the trematode's northernmost limit was Cape Cod. But with the Gulf of Maine warming faster than any other coastal region in the world, this species appears to be on the move. What seems like instability to us is, for parasites, a wave of new frontiers.

Urbanization adds another layer of risk. The United Nations Department of Economic and Social Affairs projects that by 2050 nearly 68 percent of humanity will live in cities, many of which sprawl across fragmented landscapes where humans, wildlife, and livestock are brought into intimate contact.[4] Cities create new interfaces: rodents thriving in alleyways, bats roosting in buildings, mosquitoes breeding in rooftop gutters. These synanthropic species often serve as reservoirs for pathogens, turning urban centers into hot spots of zoonotic spillover. The COVID-19 pandemic revealed just how vulnerable dense, interconnected cities are when novel pathogens emerge. Far from being ecological voids, cities are living laboratories where parasites adapt, persist, and spill over with unnerving ease.

Our food systems amplify this problem. Industrial agriculture and aquaculture concentrate vast numbers of genetically similar hosts—chickens in broiler houses, salmon in sea pens, oysters in coastal farms. For parasites, these are banquet halls. Sea lice spread explosively in crowded salmon farms before leaking into wild populations. Poultry farms foster respiratory viruses that can evolve into pandemic strains. Shrimp ponds and oyster beds, stripped of ecological complexity, become experimental arenas for parasitic outbreaks at scales nature never imagined. The systems we rely on for sustenance have become the crucibles of new parasitic challenges.

In response, the scientific community has championed the idea of One Health: the recognition that human health, animal health, and environmental health are inseparable.[5] It is an appealing, almost self-evident framework—how could one separate the mosquito from the wetland, or the virus from the bat it inhabits? One Health encourages collaboration across medicine, veterinary science, and ecology to anticipate and prevent spillovers before they happen. At its best, it reminds us that we cannot silo our defenses against disease in hospital wards alone; the battle must be waged in forests, farms, and cities alike.

Yet One Health is not without criticism. For some, it risks becoming more slogan than solution. It is often easier to declare the interconnectedness of health than to act on it meaningfully, especially when actions run against entrenched economic interests. Industrial farming continues, and will continue largely unabated, even as we acknowledge its role in disease emergence. In the United States, the lobbyists who vouch for these companies are robust and well-paid. Urban planning rarely integrates disease ecology into its designs, despite abundant evidence that the shape of green and gray spaces influences zoonotic hazards. Too often, One Health plays out as an exercise in monitoring and reacting: tracking pathogens, building diagnostic networks, preparing emergency responses. What it tends to neglect are the root conditions that make outbreaks likely in the first place: poverty, inequality, deforestation, industrial farming, and the unraveling of ecosystems.

Take, for example, a family living on the margins of a fast-growing city in West Africa. Their small house backs onto a patch of scrub forest, where bats roost in the trees. Every evening the bats spill out, flying over open drainage ditches that run past their neighborhood. The father makes his living driving a motorcycle taxi, often ferrying workers to and from a nearby poultry farm. The mother sells vegetables in a crowded open-air market where chickens, goats, and bushmeat carcasses are all handled in the same space. Their children play barefoot in alleys where water stagnates after rains, breeding clouds of mosquitoes. From the perspective of One Health, this is a textbook case of risk: wildlife, livestock, and people crammed into a web of interactions that make the spillover of parasites and pathogens almost inevitable. Yet the people living in such landscapes are rarely passive victims of disease ecology. In many parasite-endemic regions, communities possess deep practical knowledge about the organisms that share their environment. Farmers learn which seasons bring surges of mosquitoes, fishers recognize the signs of parasitic infection in their catch, and pastoralists know when livestock are most vulnerable to ticks and worms. This knowledge may not always be framed in the language of microbiology or epidemiology, but it reflects generations of close observation and adaptation. Ironically, it is precisely this type of locally grounded expertise that is often overlooked when disease interventions are designed from distant laboratories or policy offices.

What programs typically reach this family? There might be a pilot project that trains clinicians to recognize unusual fevers, or a mobile diagnostic unit that samples chickens for avian flu. These interventions matter, but they arrive after exposure has already happened. What the family really needs is harder: clean water, safe housing, health-care access, and an economy that does not push them into daily contact with risky interfaces. Until those structural conditions change, every new diagnostic test or outbreak report is just a bandage on a wound that never closes. Without grappling with these root causes, One Health risks papering over systemic vulnerabilities while leaving the underlying machinery of emergence intact.

The Anthropocene is, then, a paradoxical stage. It is an age of ecological unraveling, but also of ecological invention—an age in which parasites are poised to exploit every gap we create. Climate change redraws their maps, cities multiply their hosts, food systems concentrate their prey, and global networks transport them farther and faster than evolution alone could have managed. One Health offers a hopeful way forward, but only if it moves beyond rhetoric and grapples with the hard political and ecological realities that drive parasitism in a human-dominated world.

Parasites have always adapted to the worlds their hosts build around them. The question for the Anthropocene is whether we will learn to anticipate these adaptations—or whether, once again, we will only recognize their ingenuity after it is too late.

PLAYING GOD WITH PARASITES

In 1975, David Cronenberg released *Shivers*, a low-budget Canadian horror film that became infamous for its grotesque premise. The story begins in a sterile high-rise apartment complex, a supposed utopia of modern living. But inside its walls, a scientist has been conducting secret experiments—designing a parasite meant to replace damaged human organs. The experiment goes wrong. The parasite mutates into something new: a writhing, sluglike creature that enters through the mouth and turns its hosts into hypersexual predators. Containment fails. The infection spreads floor by floor until the entire building is overrun, its residents no longer themselves but vessels for the parasite's will. Even in an age in which schlocky exploitation films were a dime a dozen, critics were horrified; some called it "the most repulsive movie I've ever seen." It also resulted in Cronenberg being kicked out of his apartment in Toronto because of a "morality clause" in his lease that was apparently violated by this movie. Yet beneath the body horror lay a prescient question: What happens when human ambition collides with ancient, adaptable strategies of parasites? Cronenberg imagined a lab-born organism that escaped into society, exploiting the very intimacy that makes us human. Nearly fifty years later, the speculative edge of these types of movies feels less like pulp fantasy and more like a warning of the biotech age we now inhabit.

Today, in laboratories around the world, scientists wield tools of extreme precision—genetic scissors that can snip and replace DNA, synthetic biology platforms that can rewrite genomes, and artificial intelligence algorithms that can predict spillovers before they occur. Parasites, once nature's own genetic engineers, are now subject to our tinkering. We can reprogram them as medicine or model organisms, but when we do, the question is: Will they remain tools, or will they become threats?

Synthetic biology and genetic engineering now give researchers the ability to disarm or retool parasites in ways once unimaginable. In medicine, some scientists are exploring engineered helminths to treat autoimmune disorders.[6] This work is based on the hygiene hypothesis: that our immune systems coevolved with parasites, and their absence in industrialized societies may underlie the surge in allergies and inflammatory disease. Reintroducing carefully modified worms could restore balance—not as invaders, but as healers—or so these researchers believe. Elsewhere, parasites are being "tamed" into vaccines. Genetic engineering has yielded live-attenuated *Leishmania* designed to persist just enough to stimulate immunity without causing disease.[7] Similar strategies are being applied to malaria, where genetically attenuated *Plasmodium* spores can be prevented from transforming in the liver, providing sterilizing immunity in trials.[8] What were once regarded as scourges of humanity may yet become shields. Bacteriophages—viruses that prey on bacteria—are being revived as weapons against antibiotic-resistant infections. By tweaking their genomes, phages could be tuned into surgical instruments that kill specific bacterial strains no antibiotic can touch. Parasites may even serve as chassis for new therapies, their genetic plasticity repurposed for medicine rather than misery.

Artificial intelligence, too, offers a bright vision. Machine learning can now scan enormous datasets, including climate trends, migration patterns, and genomic sequences, to predict where and when spillovers are likely. In principle, AI could flag hot spots before the first case appears, giving humanity time to act. Gene drives, meanwhile, offer perhaps the boldest dream of all: By biasing inheritance, engineering tools like CRISPR-Cas9 could spread engineered traits through entire populations at lightning speed.[9] Some envision using them to collapse malaria-carrying mosquitoes altogether, erasing one of humanity's oldest parasitic scourges.

But every promise casts a shadow. The tools that let scientists disarm parasites can, in the wrong hands, be turned to arm them—editing genomes to broaden host range, evade detection, or amplify harm. Sometimes referred to as dual-use risks, this is a key point of concern for those involved in national and international security. Unlike nuclear programs, which demand vast infrastructure, many biotechnologies are comparatively cheap and distributed: CRISPR reagents, DNA synthesis services, and cloud-based bioinformatics are widely accessible. The barrier to entry has never been lower.

The fragility is not merely theoretical. Look back to a single, startling precedent: Eric Kranz, a parasitology postdoctoral student in Montreal in 1970, allegedly used *Ascaris* eggs to sicken his roommates after a dispute.[10] This case involved no CRISPR, no automation—just access, intent, and biological material. Two of the victims nearly died. Kranz's case is sobering because it shows how slim the ingredients for harm can be when motive and opportunity align. Now imagine the same motive transposed into an era of open data, lab automation, and powerful AI.

Picture this: A small unaffiliated group harvests public climate and trade datasets, combines them with open genomic repositories, and leans on widely accessible AI tools to synthesize hypotheses about fragile food systems. The AI models all point to a weakness: a parasite that thrives in a specific aquaculture practice, spreads slowly enough to evade immediate detection, and causes economic collapse in coastal communities without catastrophic human mortality. The group, lacking formal credentials, uses off-the-shelf gene-editing reagents and cloud-managed lab automation to prototype a modified strain in a clandestine setting; they outsource innocuous-looking synthesis and rely on global shipping to seed a few upstream farms. The first signal is scattered die-offs of fish and oysters in remote ponds; by the time surveillance picks up a pattern, markets have reacted—trade bans, price spikes, and panic—while investigators argue over whether the event was natural, accidental, or deliberate.

The point of this fictional vignette is not to alarm but illuminate a structural truth: The combination of AI, democratized biotechnology, and porous governance creates new pathways from motive to impact. Whereas Kranz required physical access and tacit skill, future misuse could leverage automation and algorithmic synthesis to close the gap between intent and large-scale harm. To drive this point home, consider a recent case in which non-STEM students at the Massachusetts Institute of Technology were asked to investigate whether large language models (specifically ChatGPT 3.5 and 4) could be goaded into providing crucial information for engineering a pandemic.[11] One would think such a request would be denied or flagged at the very least. Instead, the chatbot not only provided a candidate list of potential pathogens that could be used to initiate a pandemic, it explained in detail how they could be produced using synthetic DNA, provided names of biotech companies that are unlikely to screen orders, and outlined protocols and troubleshooting tips. In effect, it provided nonspecialists with a rough handbook for generating a pandemic. The issue has become so concerning that it was extensively covered in a 2023 U.S. Congressional Research Service Report.[12]

The key lesson is structural, not technical: AI and automation can deskill much of the tacit expertise that once acted as a gatekeeper. Tools that summarize decades of literature, propose experimental directions, or automate routine protocols lower the threshold from a handful of expert labs to a much broader set of actors. A system designed to forecast spillovers can, in the wrong hands, identify the most effective targets and pathways for disruption. AI can be a shield—but it can also be a sword.

It is here that speculation grows unnerving. The anthrax mailings of 2001 showed how crude dissemination of a pathogen can inflict fear, economic disruption, and political chaos. With today's tools, the conceivable targets broaden. One can imagine engineered *Toxoplasma* strains that more strongly alter behavior, microsporidia designed to undermine crop or aquaculture yields, or protozoans modified to persist in supply chains. These scenarios remain hypothetical—but the narrowing gap between imagination and feasibility demands attention.

I want to be careful in how I talk about this: Thoughtful, evidence-based analysis does not mean providing instructions. The point is to illuminate risks so policy and practice can reduce them—by hardening surveillance, tightening governance around sensitive materials and synthesis services, improving attribution capability, and investing in resilient food and health systems. As philosopher Toby Ord warns in *The Precipice*, biotechnology and artificial intelligence are among the century's highest-impact risks precisely because human-directed changes can be fast, global, and catastrophic if misused. The ethical challenge is stark: How do we reap the lifesaving benefits of genomic science, AI, and automation without opening new pathways for harm?

What parasites teach us is that control is never absolute. They thrive in blind spots, exploiting cracks in defenses and unexpected niches. Our technologies are no different. The challenge is not to eliminate risk but to anticipate where blind spots will emerge and to design systems resilient enough to absorb surprises. This means embedding ethical foresight into research, building norms of responsibility, and treating governance not as an afterthought but as part of the science. Exactly how this should be translated into an actionable framework lies beyond the scope of this book, but it is certainly an issue that must be addressed by policy and risk experts if we as a collective choose to wade into these unchartered waters.

One final reflection: Technology magnifies what parasites have always embodied—the thin line between dependence, control, and destruction. Just as parasites evolve to take advantage of host vulnerabilities, our innovations create vulnerabilities of their own—ecological, social, and ethical blind spots that we often notice only when it is too late. The question is not whether parasites will continue to shape our world; they will. The question is whether, in our pursuit of mastery, we will unwittingly become their engineers.

ANOTHER CALL FOR PARASITE CONSERVATION

It is tempting to imagine a world without parasites—no malaria, no tapeworms, no blood flukes or botflies. This is the goal of many nonprofit agencies and global health partnerships working across malaria-endemic regions. The scale of that effort is enormous. For example, the Bill & Melinda Gates Foundation has invested heavily in malaria control, research, and eradication efforts, both directly and through the Global Fund to Fight AIDS, Tuberculosis and Malaria. By the 2023–2025 replenishment period, the foundation's cumulative contribution to the Global Fund had reached $3.92 billion, and in 2025 it pledged an additional $912 million for the 2026–2028 cycle. The Global Fund itself had invested more than $20.3 billion in malaria control programs by June 2025.[13] These numbers demonstrate the scale of philanthropic investment in parasite eradication. On the surface, it sounds like paradise: ecosystems freed from invisible saboteurs, humanity spared from billions of infections each year. But evolution rarely writes stories with such simple morals. To remove parasites wholesale would not be a cure—it would be an unraveling.

As a thought experiment, imagine that overnight every parasite on Earth vanished. At first the consequences might look like cause for celebration. Farmers would wake to oyster beds free of mud blister worms, herds free of ticks, fields spared from fungal rusts. Hospitals would be empty of malaria patients; veterinarians would find their caseloads cut in half. Humanity would cheer.

But then the cracks would appear.

Predator-prey dynamics, once stabilized by parasitic checks, would lurch out of balance. Herbivores spared from nematodes would boom, stripping landscapes bare. Fisheries without parasitic regulators would swing wildly between gluts and collapses. Ecosystems that had coevolved with parasites for millennia would lose the invisible threads that held them together. If you are approaching the end of this book, you know that parasites are the governors of biodiversity. Remove them, and whole networks of life begin to wobble. To borrow an analogy from my own field of marine science, they are like the deep-sea conveyor belts of the planet's oceans—the thermohaline currents that shuttle heat, salt, and nutrients around the globe. Most of the time, we do not notice these invisible flows, but they keep climates stable, fisheries productive, and ecosystems balanced. Shut down even part of that conveyor—say, by freshwater flooding into the North Atlantic—and the entire system buckles, triggering climate chaos on distant continents. Parasites operate the same way in ecological systems—unseen regulators whose absence could send unpredictable shocks rippling outward, destabilizing communities in ways we barely understand.

The effects would reach into our own bodies. For most of our evolutionary history, immune systems have developed in constant dialogue with parasites. Helminths and protozoans were as much teachers as threats, calibrating the balance between aggression and tolerance. Their absence has already been linked to rising autoimmune diseases and allergies in industrialized societies. If every parasite were gone, the hygiene hypothesis would no longer be a hypothesis—it would become reality. Immune systems would drift without their ancient sparring partners, overreacting to harmless triggers, turning against their own tissues. The victory over parasites might come at the cost of epidemics of asthma, Crohn's disease, and lupus.

The larger point is this: Parasites are stabilizers as well as destroyers. They shape host populations, enforce limits, and sustain diversity by preventing dominance. Their presence can be cruel at the level of individuals, but at the scale of ecosystems they are often the hidden architects of balance.

Parasite conservation, then, is not about the sentimentality for tapeworms or mosquitoes; it is about humility. Life has woven parasites into every habitat, every food web, every body. To imagine biology without them is to imagine a world both less stable and less rich. As discussed earlier in the book, parasitism was likely a key feature of the emergence of life, and its subsequent evolution across the gnarly branches on the tree of life is testament to its permanence on our planet. The case for their conservation is ultimately the case for recognizing that the things we fear most may also be the things that keep us alive.

History gives us glimpses of what happens when parasites are erased. When the cattle plague rinderpest was eliminated from Africa in the twentieth century, wildebeest and buffalo populations exploded. At first this seemed like triumph, but the boom transformed savannas: Grasses were grazed down, fires became less frequent, and dense woodlands spread.[14] Lion numbers rose, but so did conflicts with people as habitats shifted. A single parasite's disappearance cascaded into continent-scale ecological change. If the disappearance of a single pathogen can reshape landscapes, imagine the effect of losing them all.

Some might argue that surely not *all* parasites matter. What about those that rely solely on humans? Consider the intestinal roundworm or the Guinea worm. They cycle exclusively through us as final hosts, bringing pain, disability, or death, yet playing no obvious role in broader ecosystems. Would their surgical elimination be not only harmless but humane? The global campaign against Guinea worm, funded in large part by the Carter Center and international partners, is celebrated precisely because it targets a parasite whose disappearance seems to carry no ecological cost. The argument is powerful.

And yet, the story does not end there. For one thing, the ecological isolation of a parasite is rarely as clear-cut as it seems. Hosts we once thought were exclusive often turn out to have hidden reservoirs, and parasites we believed to be evolutionary dead ends can surprise us with unexpected connections. Even when they are genuinely confined to humans, their role as selective pressures remains. Our immune systems are not static—they are sculpted by constant dialogue with parasites. Remove them, and you may lift a burden, but you may also shift the evolutionary pressures in ways we cannot predict. The rise of autoimmune and allergic diseases in societies that have effectively purged helminths from daily life is a cautionary reminder that absence, too, can be pathogenic.

Finally, there is the ethical dimension. To deliberately erase a species—however grotesque—forces us into the role of editors of evolution. Do we have the right to decide that a life-form's only legacy is to vanish? To some, parasites like *Plasmodium*, *Ascaris*, or Guinea worm embody nothing but suffering, and their eradication is moral progress. To others, their elimination, however pragmatic, is another example of human hubris: assuming we can prune the tree of life without unintended consequences.

THE ARCHITECTS REMAIN

If conservation forces us to imagine what might be lost without parasites, extinction crises also push us to ask a harder question: What comes after? If we succeed in protecting biodiversity, or even if we fail, evolution will not stop. Parasites, with their small genomes, rapid cycles, and uncanny ability to adapt, are not likely to be the first casualties of upheaval. More often, they are the stragglers, the opportunists, the survivors that slip through when others fall away. To think about the future of life on Earth means reckoning not only with which species vanish but with which strategies endure. And parasitism, for all its costs, is among the most enduring strategies evolution has ever produced.

If you strip life down to its most durable blueprints, parasitism will be among the last left standing. Parasites do not require vast habitats, towering bodies, or intricate societies. They thrive tucked into overlooked crevices, hidden in tissues, or suspended in microscopic cysts waiting for the right hosts to appear. They are masters of persistence, passed forward in the bloodstream, the soil, and the sea spray.

History suggests that when ecosystems collapse, parasites often endure. After mass extinctions, paleontologists find their traces not in grand fossils, but in the scars left behind on bones and shells—boreholes from ancient worms, cysts in dinosaur vertebrae, eggs trapped in amber. When giant lineages vanished, parasites quietly adapted, shifting hosts, reshaping cycles, and waiting out catastrophe.

Now imagine the Anthropocene ending not with human triumph but with human absence. Cities crumble, farmland reverts to scrub, climate chaos redraws the map of habitability. What survives? Rats and cockroaches, pigeons and feral dogs—and the parasites that ride along with them. The intestinal worms of humans slip easily into survivors. The fleas and ticks that once plagued us attach to whatever mammals remain. Even the viruses embedded in our DNA—endogenous retroviruses, evolutionary hitchhikers buried in our genomes—will not vanish with us. They will be carried forward, stitched into the blueprints of whatever species inherits the Earth. Parasitism does not end with us; it simply finds new stages. Perhaps a world dominated by insects, as many biologists predict, will also be a world dominated by insect parasites—fungi that turn ants into zombies, wasps that weaponize caterpillars into bodyguards. Perhaps in warming oceans, parasitic copepods and isopods will proliferate, sculpting fish populations long after our fleets are gone.

In this sense, parasites are the custodians of evolution's future. Long after languages fade and skyscrapers sink into the sea, parasites will still be practicing their ancient art of survival, shaping, pruning, and redirecting the course of whatever life remains. They are, in the end, the truest heirs of Earth's resilience.

I return, in memory, to my first semester teaching about parasites at Wheaton College—the day I dimmed the lights, pressed play, and watched the room recoil as a surgeon lifted a writhing tangle of worms from a patient's abdomen. Gasps of disbelief. For a moment, the students were horrified. And then, almost imperceptibly, they leaned forward. Fascination overtook revulsion, and the room filled with questions, ideas, and awe. The parasites had done what no lecture could: they had seized the strings. When all is said and done, that is their enduring power. Parasites unsettle us, but they also reveal. They strip away our illusions of separateness, forcing us to confront the truth that life is never truly solitary, that every creature is bound to another in webs of dependence, struggle, and creativity. They are grotesque, yes, but also profoundly instructive—nature's most unsettling teachers. As we look to the future, parasites will remain with us. They will spark epidemics, recalibrate ecosystems, and etch their mark into genomes yet unborn. They will continue to pull at the threads of human history, sometimes as destroyers, sometimes as innovators. In their reflection, we glimpse both the fragility of our place on Earth and the resilience of life itself.

When my students left that classroom, they carried more than a memory of disgust. They glimpsed a deeper dimension of life. In the end, the choice is ours. We can resist the parasite's strings, clinging to the fantasy of human mastery; or we can learn from them—seeing vulnerability not as a weakness but as a connection, seeing dependence not as failure but as the engine of life. The worms on that screen, the pathogens in our blood, the unseen architects in our DNA—they are telling us something. They are asking us to decide what lessons we will carry forward.

NOTES

INTRODUCTION

1. S. B. Weinstein and A. M. Kuris, "Independent Origins of Parasitism in Animalia," *Biology Letters* 12 (2016): 20160324.
2. F. Xu et al., "On the Reversibility of Parasitism: Adaptation to a Free-Living Lifestyle via Gene Acquisitions in the Diplomonad *Trepomonas* sp. PC1," *BMC Biology* 14, 62 (2016), https://doi.org/10.1186/s12915-016-0284-z.
3. J. Klunk et al., "Evolution of Immune Genes Is Associated with the Black Death," *Nature* 611 (2022): 312–319, https://doi.org/10.1038/s41586-022-05349-x.
4. K. D. Lafferty et al., "Parasites Dominate Food Web Links," *Proceedings of the National Academy of Sciences of the United States of America* 103, no. 30 (2006): 11211–11216, https://doi.org/10.1073/pnas.0604755103.

1. PARASITES AS EVOLUTION'S ENGINEERS

1. Centers for Disease Control and Prevention (CDC), "Clinical Features of Zoonotic Hookworm," February 15, 2024, accessed May 11, 2026, https://www.cdc.gov/zoonotic-hookworm/hcp/clinical-features/index.html.
2. The Nobel Prize, "Nobel Prize in Physiology or Medicine 2015," accessed October 12, 2025, https://www.nobelprize.org/prizes/medicine/2015/summary/.
3. A. D. Santin et al., "Ivermectin: A Multifaceted Drug of Nobel Prize–Honoured Distinction with Indicated Efficacy Against a New Global Scourge, COVID-19," *New Microbes and New Infections* 43 (2021): 100924, https://doi.org/10.1016/j.nmni.2021.100924.

4. Video clip from William Campbell's Nobel Prize Banquet speech, accessed October 12, 2025, https://www.youtube.com/watch?v=knLelJU-wRU.
5. Quoted in ecropolis, "Parasitology: Picking Up Where Medical School Left Off," Parasites Without Borders, August 5, 2024, accessed October 12, 2025, https://parasiteswithoutborders.com/parasitology-picking-up-where-medical-school-left-off/.
6. A. A. David, "A Student-Centered Framework for Teaching Undergraduate Parasitology," *Trends in Parasitology* 33, no. 6 (2017): 420–423, https://doi.org/10.1016/j.pt.2017.01.010.
7. A. Shetty et al., "Medical Symbols in Practice: Myths vs Reality," *Journal of Clinical and Diagnostic Research* 8, no. 8 (2014): PC12–PC14, https://doi.org/10.7860/JCDR/2014/10029.4730.
8. C. Strube et al., "Simple, but Not Easy—Opportunities and Challenges from Teachers' and Students' Perspectives in the 21st Century of Veterinary Parasitology Teaching," *Veterinary Parasitology* 252 (2018): 74–79, https://doi.org/10.1016/j.vetpar.2018.01.034.
9. H. Hooshyar and P. Rostamkhani, "Accurate Laboratory Diagnosis of Human Intestinal and Extra-Intestinal Amoebiasis," *Gastroenterology and Hepatology from Bed to Bench* 15, no. 4 (2022): 343–359, https://doi.org/10.22037/ghfbb.v15i4.2496.
10. Centers for Disease Control and Prevention (CDC), "DPDx—Laboratory Identification of Parasites of Public Health Concern: Taeniasis," last reviewed June 7, 2024, accessed October 12, 2025.
11. K. S. Warren and A. A. F. Mahmoud, "Tapeworms," in *Geographic Medicine for the Practitioner*, ed. K. S. Warren and A. A. F. Mahmoud (Springer, 1985), 182–189.
12. A. Dobson et al., "Homage to Linnaeus: How Many Parasites? How Many Hosts?" *Proceedings of the National Academy of Sciences of the United States of America* 105, no. 1 (2008): 11482–11489, https://doi.org/10.1073/pnas.0803232105; R. F. Hechinger, "Parasites Help Find Universal Ecological Rules," *Proceedings of the National Academy of Sciences of the United States of America* 112, no. 6 (2015): 1656–1657, https://doi.org/10.1073/pnas.1423785112.
13. L. Van Valen, "A New Evolutionary Law," *Evolutionary Theory* 1 (1973): 1–30.
14. J. Rolff, "Why Did the Acquired Immune System of Vertebrates Evolve?," *Developmental & Comparative Immunology* 31 (2007): 476–482.

15. C. Hammoud et al., "Trematode Infection Affects Shell Shape and Size in *Bulinus tropicus*," *International Journal for Parasitology: Parasites and Wildlife* 18 (2022): 300–311.
16. A. Kajimoto et al., "Morphological Feminization in Hermit Crabs (Family Paguridae) Induced by Rhizocephalan Barnacles," *Zoological Letters* 11, no. 1 (2025): 6, https://link.springer.com/article/10.1186/s40851-025-00252-5.
17. W. Souder, "In Minnesota Lakes, an Alarming Mystery," *Washington Post*, September 29, 1996.
18. T. Lyden, "Un-Frog-getable: What Happened to Minnesota's Deformed Frogs?," FOX 9 (Minneapolis–St. Paul), updated May 10, 2021, https://www.fox9.com/news/un-frog-getable-what-happened-to-minnesotas-deformed-frogs.
19. J. Helgen, *Peril in the Ponds: Deformed Frogs, Politics, and a Biologist's Quest* (University of Massachusetts Press, 2012).
20. P. T. J. Johnson et al., "The Effect of Trematode Infection on Amphibian Limb Development and Survivorship," *Science* 284, no. 5415 (1999): 802–804, https://doi.org/10.1126/science.284.5415.802.
21. A. Muehlenbachs et al., "Malignant Transformation of *Hymenolepis nana* in a Human Host," *New England Journal of Medicine* 373, no. 19 (2015): 1845–1852, https://doi.org/10.1056/NEJMoa1505892.
22. S. G. Sanchez and S. Besteiro, "The Pathogenicity and Virulence of *Toxoplasma gondii*," *Virulence* 12, no. 1 (2021): 3095–3114, https://doi.org/10.1080/21505594.2021.2012346.
23. P. R. Torgerson and P. Mastroiacovo, "The Global Burden of Congenital Toxoplasmosis: A Systematic Review," *Bulletin of the World Health Organization* 91, no. 7 (2013): 501–508, https://doi.org/10.2471/BLT.12.111732.
24. L. Bollani et al., "Congenital Toxoplasmosis: The State of the Art," *Frontiers in Pediatrics* 10 (2022), https://doi.org/10.3389/fped.2022.894573.
25. K. Sharun et al., "SARS-CoV-2 in Animals: Potential for Unknown Reservoir Hosts and Public Health Implications," *Veterinary Quarterly* 41, no. 1 (2021): 181–201, https://doi.org/10.1080/01652176.2021.1921311.
26. O. Simonetti et al., "Dracunculiasis Over the Centuries: The History of a Parasite Unfamiliar to the West," *Infezioni in Medicina* 31, no. 2 (2023): 257–264, https://www.infezmed.it/media/journal/Vol_31_2_2023_15.pdf.
27. Carter Center, "Guinea Worm Case Totals," accessed October 13, 2025, https://www.cartercenter.org/health/guinea_worm/case-totals.html.

28. N. St. Fleur, "The Second Human Disease Ever Eradicated," *Atlantic*, January 15, 2015, accessed October 13, 2025, https://www.theatlantic.com/health/archive/2015/01/carter-center-guinea-worm-jimmy-carter-parasite-pipe-filter/384557/.
29. R. A. McDonald et al., "Ecology of Domestic Dogs *Canis familiaris* as an Emerging Reservoir of Guinea Worm *Dracunculus medinensis* Infection," *PLoS Neglected Tropical Diseases* 14, no. 4 (2020): e0008170, https://doi.org/10.1371/journal.pntd.0008170.
30. M. Milinski and T. C. M. Bakker, "Female Sticklebacks Use Male Coloration in Mate Choice and Hence Avoid Parasitized Male," *Nature* 344 (1990): 330–333.
31. Wetland and Aquatic Research Center, "Invasive Burmese Pythons in Southern Florida," U.S. Geological Survey, September 30, 2024, accessed October 13, 2025, https://www.usgs.gov/centers/wetland-and-aquatic-research-center/science/invasive-burmese-pythons-southern-florida.
32. D. C. Card et al., "Novel Ecological and Climatic Conditions Drive Rapid Adaptation in Invasive Florida Burmese Pythons," *Molecular Ecology* 27 (2018): 4744–4757.
33. U.S. Geological Survey, "Diet of Invasive Burmese Pythons (*Python molorus bivittatus*) in Southern Florida, 1995–2020 [data set], July 10, 2023, accessed October 13, 2025, https://www.usgs.gov/data/diet-invasive-burmese-pythons-python-molorus-bivittatus-southern-florida-1995-2020.
34. M. A. Miller et al., "Parasite Spillover: Indirect Effects of Invasive Burmese Python," *Ecology and Evolution* 8 (2017): 830–840.
35. J. A. Dunne et al., "Parasites Affect Food Web Structure Primarily Through Increased Diversity and Complexity," *PLoS Biology* 11 (2013): e1001579.

2. THE GENOMIC BATTLEFIELD

1. World Health Organization, "Trypanosomiasis, Human African (Sleeping Sickness)," May 2, 2023, retrieved October 13, 2025, https://www.who.int/news-room/fact-sheets/detail/trypanosomiasis-human-african-(sleeping-sickness).
2. J. D. Watson and F. H. C. Crick, "Molecular Structure of Nucleic Acids: A Structure for Deoxyribose Nucleic Acid," *Nature* 171 (1953): 737–738.

3. W. Harnett and M. M. Harnett, "Epigenetic Changes Induced by Parasitic Worms and Their Excretory-Secretory Products," *Biochemical Society Transactions* 52 (2024): 55–63.
4. J. Mao et al., "Human Endogenous Retroviruses in Development and Disease," *Computational and Structural Biotechnology Journal* 19 (2021): 5978–5986.
5. J. Wang et al., "Endogenous Retroviruses in Development and Health," *Trends in Microbiology* 32 (2024): P342–P354.
6. A. Hayward et al., "Pan-Vertebrate Comparative Genomics Unmasks Retrovirus Macroevolution," *Proceedings of the National Academy of Sciences of the USA* 112 (2015): 464–469.
7. C. Lavialle et al., "Paleovirology of "Syncytins," Retroviral Env Genes Exapted for a Role in Placentation, *Philosophical Transactions of the Royal Society B* 368 (2013).
8. A. P. Jason de Koning et al., "Repetitive Elements May Comprise Over Two-Thirds of the Human Genome," *PLoS Genetics* 7 (2011).
9. J. Radwan et al., "Advances in the Evolutionary Understanding of *MHC* Polymorphism," *Trends in Genetics* 36 (2020): 298–311.
10. L. Luzzatto et al., "Glucose-6-phosphate Dehydrogenase Deficiency," *Blood* 136 (2020): 1225–1240.
11. H. Quach et al., "Different Selective Pressure Shape the Evolution of Toll-like Receptors in Human and African Great Ape Populations," *Human Molecular Genetics* 22 (2013): 4829–4840.
12. E. M. Mewamba et al., "The Genetics of Human Schistosomiasis Infection Intensity and Liver Disease: A Review," *Frontiers in Immunology* 12 (2021).
13. A. V. Nguyen and A. M. Soulika, "The Dynamics of the Skin's Immune System," *International Journal of Molecular Sciences* 20 (2019): 1811.
14. M. Pagel and W. Bodmer, "A Naked Ape Would Have Fewer Parasites," *Proceedings of the Royal Society of London* B 270 (2003): S117–S119.
15. Y. Zou et al., "Helminth Reshapes Host Gut Microbiota and Immunoregulation by Deploying an Antimicrobial Program of Innate Immunity," *Gut Microbes* 17 (2025): 2496447.
16. L. T. Stiemsma et al., "The Hygiene Hypothesis: Current Perspectives and Future Therapies," *ImmunoTargets and Therapy* (2015): 143–157.

3. THE GREAT GENETIC HEIST

1. S. M. Soucy et al., "Horizontal Gene Transfer: Building the Web of Life," *Nature Reviews Genetics* 16 (2015): 472–482.
2. CBS News, "Woman Says Her Son Couldn't Afford His Insulin—Now He's Dead," January 4, 2019, https://www.cbsnews.com/news/mother-fights-for-lower-insulin-prices-after-sons-tragic-death/.
3. G. A. Petsko, "Transformation," *Genome Biology* 7 (2006): 117.
4. National Human Genome Research Institute, "1944: DNA Is 'Transforming Principle,'" retrieved October 13, 2025, https://www.genome.gov/25520250/online-education-kit-1944-dna-is-transforming-principle.
5. U.S. Food and Drug Administration, "100 Years of Insulin," accessed October 13, 2025, https://www.fda.gov/about-fda/fda-history-exhibits/100-years-insulin.
6. J. A. Lindsay, "*Staphylococcus aureus* Genomics and the Impact of Horizontal Gene Transfer," *International Journal of Medical Microbiology* 304 (2014): 103–109.
7. A. G. Kent et al., "Widespread Transfer of Mobile Antibiotic Resistance Genes Within Individual Gut Microbiomes Revealed Through Bacterial Hi-C," *Nature Communications* 11 (2020): 4379.
8. Center for Disease Control and Prevention (CDC), "About the National Antimicrobial Resistance Monitoring Systems (NARMS)," September 6, 2024, https://www.cdc.gov/narms/about/index.html.
9. National Intelligence Council, "Updated Assessment on COVID-19 Origins," based on information through August 2021, https://www.dni.gov/files/ODNI/documents/assessments/Declassified-Assessment-on-COVID-19-Origins.pdf.
10. S. K. Garushyants et al., "Template Switching and Duplications in SARS-CoV-2 Genomes Give Rise to Insertion Variants That Merit Monitoring," *Communication Biology* 4 (2021): 1343.
11. E. Sallard et al., "Tracing the Origins of SARS-COV-2 in Coronavirus Phylogenies: A Review," *Environmental Chemistry Letters* 19 (2021): 769–785.
12. J. Deen et al., "Epidemiology of Cholera," *Vaccine* 38 (2020): A31–A40; V. Mishra, "Yemen Bears World's Highest Cholera Burden, Deepening Humanitarian Crisis," *UN News*, December 23, 2034, https://news.un.org/en/story/2024/12/1158491.

13. M. K. Waldor and J. J. Mekalanos, "Lysogenic Conversion by a Filamentous Phage Encoding Cholera Toxin," *Science* 272 (1996): 1910–1914.
14. A. Nasir and G. Caetano-Anolles, "A Phylogenomic Data-Driven Exploration of Viral Origins and Evolution," *Science Advances* 1 (2015): e1500527.
15. S. Sun et al., "Structural Studies of the Sputnik Virophage," *Journal of Virology* 84 (2010): 894–897.
16. C. R. Darwin, letter to A. Gray, May 22, 1860, in *The Correspondence of Charles Darwin*, Vol. 8 (1860), ed. F. Burkhardt et al. (Cambridge University Press, 2023), https://doi.org/10.1017/9781009233606.
17. S. LaFee and H. Buschman, "Novel Phage Therapy Saves Patient with Multidrug-Resistant Bacterial Infection," *UC San Diego Today*, April 25, 2017, https://today.ucsd.edu/story/novel_phage_therapy_saves_patient_with_multidrug_resistant_bacterial_infect.
18. D. Zarlenga et al., "Horizontal Gene Transfer Provides Insights Into the Deep Evolutionary History and Biology of *Trichinella*, *Food and Waterborne Parasitology* 27 (2022): e00155.
19. Al Jazeera English, "Lebanon: Fears as 'Aleppo Boil' Disease Spreads," YouTube, https://www.youtube.com/watch?v=xp3VL7lORGo.
20. D. Steverding, "The History of Leishmaniasis," *Parasites & Vectors* 10 (2017): 82.
21. G-FINDER, "Smart Decisions: The 2024 G-FINDER 2024 Neglected Disease R&D Report Executive Summary," January 30, 2025, https://cdn.impactglobalhealth.org/media/G-FINDER%202024_Executive%20summary.pdf.
22. N. Aronson et al., "Diagnosis and Treatment of Leishmaniasis: Clinical Practice Guidelines by the Infectious Diseases Society of America (IDSA) and the American Society of Tropical Medicine and Hygiene (ASTMH)," *Clinical Infectious Diseases* 63 (2016): 1539–1537.
23. A. Clementi et al., "Renal Involvement in Leishmaniasis: A Review of the Literature," *NDT Plus* 4 (2011): 147–152.
24. T. P. C. Dorlo et al., "Miltefosine: A Review of Its Pharmacology and Therapeutic Efficacy in the Treatment of Leishmaniasis," *Journal of Antimicrobial Chemotherapy* 67 (2012): 2576–2579.
25. S. Srivastava et al., "Laboratory Confirmed Miltefosine Resistant Cases of Visceral Leishmaniasis," *Parasites & Vectors* 10 (2017): 49.

26. J. A. Black et al., "Life in Plastic, It's Fantastic! How *Leishmania* Exploit Genome Instability to Shape Gene Expression," *Frontiers in Cellular and Infection Microbiology* 13 (2023): 1102462.
27. S. R. B. Uliana et al., "*Leishmania* Genomics: Where Do We Stand?," in *Bioinformatics in Tropical Disease Research: A Practical and Case-Study Approach*, ed. A. Gruber et al (Bethesda, MD: National Center for Biotechnology Information [US], 2008).
28. E. Vikeved et al., "The Dynamics of Lateral Gene Transfer in the Genus *Leishmania*—A Route for Adaptation and Species Diversification," *PLoS Neglected Tropical Diseases* 10 (2016): e0004326.
29. N. Douanne et al., "*Leishmania* Parasites Exchange Drug-Resistance Genes Through Extracellular Vesicles," *Cell Reports* 40 (2022): 111121.
30. A. N. McNulty et al., "Localization of *Wolbachia*-like Gene Transcripts and Peptides in Adult *Onchocerca flexuosa* Worms Indicates Tissue Specific Expression," *Parasites & Vectors* 6 (2013): 2.
31. K. Bell and S. R. Bordenstein, "A Margulian View of Symbiosis and Speciation: The Nasonia Wasp System," *Symbiosis* 87 (2022): 3–10.
32. Nobel Prize, "The Nobel Prize in Chemistry 2020," accessed October 24, 2025, https://www.nobelprize.org/prizes/chemistry/2020/summary/.
33. J. Kluger, "The Return of the Dire Wolf," *Time*, April 7, 2025, https://time.com/7274542/colossal-dire-wolf/. However, keep in mind that they did not really "create" a dire wolf; they genetically engineered gray wolves in such a way that the engineered traits mimicked key traits of the dire wolf, but did not produce an actual dire wolf.
34. Target Malaria, "Release of Genetically Modified Mosquitoes: A New Step in the Search for Innovative Tools to Combat Malaria in Africa," press release, Bobo-Dioulasso, August 13, 2025, https://targetmalaria.org/virtual-press-room/press-releases/release-of-genetically-modified-mosquitoes-a-new-step-in-the-search-for-innovative-tools-to-combat-malaria-in-africa/. As of September 2025, however, the trials have been suspended (https://www.isaaa.org/kc/cropbiotechupdate/article/default.asp?ID=21511).
35. Oxitec Ltd., "Groundbreaking Pilot of New Friendly™ Mosquitoes Demonstrates 96% Suppression of Dengue-Spreading Aedes aegypti in Urban Communities in Brazil," press release, October 25, 2022, https://www.oxitec.com/en/news/groundbreaking-pilot-of-new-friendly-mosquitoes.

4. MASTERS OF MANIPULATION

1. R. Dawkins, "Extended Phenotype—but Not *Too* Extended: A Reply to Laland, Turner and Jablonka," *Biology and Philosophy* 19 (2004): 377–396.
2. D. G. Biron et al., "Behavioral Manipulation in a Grasshopper Harbouring Hairworm: A Proteomics Approach," *Proceedings of the Royal Society* 272 (2005): 2117–2126.
3. T. Mishina et al., "Massive Horizontal Gene Transfer and the Evolution of Nematomorph-Driven Behavioral Manipulation of Mantids," *Current Biology* 33 (2023): 4988–4994.
4. K. Kuroda et al., "Sex-Specific Manipulation of Sexually Cannibalistic Mantid Mating Behavior by Hairworms," *Behavioral Ecology* 35 (2024): arae071.
5. M. Yang, "Family Infected with Parasitic Worms in US After Eating Bear Meat, CDC Says," *Guardian*, May 26, 2024, https://www.theguardian.com/world/article/2024/may/26/people-infected-bear-meat-parasitic-worms-trichinellosis.
6. S. Cash-Goldwasser et al., "Outbreak of Human Trichinellosis—Arizona, Minnesota and South Dakota, 2022," *Morbidity and Mortality Weekly Report* 73, no. 20 (May 24, 2024): 456–459, https://www.cdc.gov/mmwr/volumes/73/wr/mm7320a2.htm.
7. E. Pozio, "Adaptation of *Trichinella* spp. for Survival in Cold Climates," *Food and Waterborne Parasitology* 4 (2016): 4–12.
8. M. Berdoy et al., "Fatal Attraction in Rats Infected with *Toxoplasma gondii*," *Proceedings of the Royal Society of London* B 267 (2000): 1591–1594.
9. R. Calero-Bernal and S. M. Gennari, "Clinical Toxoplasmosis in Dogs and Cats: An Update," *Frontiers in Veterinary Science* 6 (2019): 54.
10. A. Bouznach et al., "Systemic *Toxoplasma gondii* Infection in a Cat with Incidental Cholangioma," *Israel Journal of Veterinary Medicine* 70 (2015): 64–67.
11. Z. T. Wang et al., "The Aromatic Amino Acid Hydroxylase Genes *AAH1 and AAH 2* in *Toxoplasma gondii* Contribute to Transmission in the Cat," *PLoS Pathogens* 13 (2017): e1006272.
12. W. H. Tong et al., "Behavioral Biology *Toxoplasma gondii* Infection," *Parasites & Vectors* 14 (2021): 77.
13. E. A. Wohlfert et al., "Brains and Brawn: *Toxoplasma* Infections of the Central Nervous System and Skeletal Muscle," *Trends in Parasitology* 33 (2017): 519–531.

14. J. Xiao et al., "Sex-Specific Changes in Gene Expression and Behavior Induced by Chronic *Toxoplasma* Infection in Mice," *Neuroscience* 206 (2012): 39–48.
15. E. A. Herniou et al., "When Parasitic Wasps Hijacked Viruses: Genomic and Functional Evolution of Polydnaviruses," *Philosophical Transactions of the Royal Society* 368 (2013): 20130051.
16. L. E. M. McMillan et al., "The Caterpillar *Manduca sexta* Brain Shows Changes in Gene Expression and Protein Abundance Correlating with Parasitic Manipulation of Behavior," *Scientific Reports* 14 (2024): 31773.
17. C. de Bekker et al., "The Ants (Hymenoptera: Formicidae) and Their Parasites: Effects of Parasitic Manipulations and Host Responses on Ant Behavioral Ecology," *Myrmecological News* 28 (2018): 1–24.
18. M. Boillat et al., "Neuroinflammation-Associated Aspecific Manipulation of Mouse Predator Fear by *Toxoplasma gondii*," *Cell Reports* 30 (2020): P320–P334.
19. R. Kopecky et al., "Le Petit Machiavellian Prince: Effects of Latent Toxoplasmosis on Political Beliefs and Values," *Evolutionary Psychology* 20 (2022): 14747049221112657.
20. C. L. Fincher and R. Thornhill, "Parasite-Stress Promotes In-Group Assortative Sociality: The Cases of Strong Family Ties and Heightened Religiosity," *Behavioral and Brain Sciences* 35 (2012): 61–79.
21. C. L. Fincher et al., "Pathogen Prevalence Predicts Human Cross-Cultural Variability in Individualism/Collectivism," *Proceedings of the Royal Society B* 275 (2008): 1279–1285.
22. G. D. A. Brown et al., "Political Attitudes and Disease Threat: Regional Pathogen Stress Is Associated with Conservative Ideology Only for Older Individuals," *Personality and Social Psychology Bulletin* 50 (2024): 1581–1595.
23. W. Jedryczka, "Can History of Parasitic Diseases Increase Social Conservatism? Testing Behavioral Immune System Theory," *Journal of Education Culture and Society* 13 (2022): 383–394.
24. D. R. Murray et al., "Pathogens and Politics: Further Evidence That Parasite Prevalence Predicts Authoritarianism," *PLoS One* 8 (2013): e62275.
25. K. Popper, *The Logic of Scientific Discovery* (Hutchinson, 1959). Originally published in German in 1934, philosopher Karl Popper argued that scientific hypotheses should be falsifiable: they should make risky predictions that could be refuted by observation or experiment.

26. P. K. Jonason and D. P. Schmitt, "Quantifying Common Criticisms of Evolutionary Psychology," *Evolutionary Psychological Science* 2 (2016): 177–188.
27. T. E. Currie and R. Mace, "Analyses Do Not Support the Parasite-Stress Theory of Human Sociality," *Behavioral and Brain Sciences* 35 (2012): 83–85.
28. K. D. Lafferty, "Can the Common Brain Parasite, *Toxoplasma gondii*, Influence Human Culture?," *Proceedings of the Royal Society* 273 (2006): 2749–2755.

5. THE HIDDEN ARMS RACE

1. D. Vergara et al., "The Geographic Mosaic of Sex and Infection in Lake Populations of a New Zealand Snail at Multiple Spatial Scales," *American Naturalist* 182 (2013): 484–492.
2. Hamilton's classic paper is W. D. Hamilton, "Sex Versus Non-Sex Versus Parasite," *Oikos* 35 (1980): 2.
3. An excellent review of antigenic variation in *P. falciparum* is A. Scherf et al., "Antigenic Variation in *Plasmodium falciparum*," *Annual Review of Microbiology* 62 (2008): 445–470.
4. A. Claessens et al., "Generation of Antigenic Diversity in *Plasmodium falciparum* by Structured Rearrangement of *Var* Genes During Mitosis," *PLoS Genetics* 10 (2014): e1004812.
5. M. Milinski, "The Function of Mate Choice in Sticklebacks: Optimizing MHC Genes," *Journal of Fish Biology* 63 (2003): 1–16.
6. D. P. L. Green, "Infectious Diseases, Imposing Density-Dependent Mortality on MHC/HLA Variation, Can Account for Balancing Selection and MHC/HLA Polymorphism," *arXiv* (2025): 2501.00767.
7. Q. Zhan et al., "Hyper-Diverse Antigenic Variation and Resilience to Transmission-Reducing Intervention in Falciparum Malaria," *Nature Communications* 15 (2024): 7343.
8. M. Recker et al., "Antigenic Variation in *Plasmodium falciparum* Malaria Involves a Highly Structured Switching Pattern," *PLoS Pathogens* 7 (2011): e1001306.
9. His seminal paper is R. E. Lenski et al., "Long-Term Experimental Evolution in *Escherichia coli*: I. Adaptation and Divergence During 2,000 Generations," *American Naturalist* 138 (1991): 1315–1341.

10. K. M. Wegner et al., "Mortality Selection During the 2003 European Heat Wave in Three-Spined Sticklebacks: Effects of Parasites and MHC Genotype," *BMC Evolutionary Biology* 8 (2008): 124.
11. C. Loiseau, "*Plasmodium relictum* Infection and MHC Diversity in the House Sparrow (*Passer domesticus*)," *Proceedings of the Royal Society* 278 (2010): 1264–1272.
12. P. J. McLaren and J. Fellay, "HIV-1 and Human Genetic Variation," *Nature Reviews Genetics* 22 (2021): 645–657.
13. Nobel Prize, "Nobel Prize in Physiology or Medicine 2025," accessed October 29, 2025, https://www.nobelprize.org/prizes/medicine/2025/summary/.
14. For a review article on the topic, see M. Hartfield and P. D. Keightley, "Current Hypotheses for the Evolution of Sex and Recombination," *Integrative Zoology* 7 (2012): 192–209.
15. In this review, 219 papers dating back to the 1950s were reviewed with respects to this topic: L. J. Buckingham and B. Ashby, "Coevolutionary Theory of Hosts and Parasites," *Journal of Evolutionary Biology* 35 (2022): 205–224.
16. C. D. Kelly et al., "Sexual Dimorphism in Immunity Across Animals: A Meta-Analysis," *Ecology Letters* 21 (2018): 1885–1894.
17. B. Goldman, "In Men, High Testosterone Can Mean Weakened Immune Response, Study Finds," Stanford Medicine New Center, December 23, 2013, https://med.stanford.edu/news/all-news/2013/12/in-men-high-testosterone-can-mean-weakened-immune-response-study-finds.html.
18. F. Angum, "The Prevalence of Autoimmune Disorders in Women: A Narrative Review," *Cureus* 12 (2020): e8094.
19. G. D. D. Hurst and C. L. Frost, "Reproductive Parasitism: Maternally Inherited Symbionts in a Biparental World," *Cold Spring Harbor Perspectives in Biology* 7 (2015): a017699.
20. M. Y. Akinyi et al., "Role of Grooming in Reducing Tick Load in Wild Baboons (*Papio cynocephalus*)," *Animal Behavior* 85 (2013): 559–568.
21. C. Poirotte et al., "Mandrills Use Olfaction to Socially Avoid Parasitized Conspecifics," *Science Advances* 3 (2017): e1601721.
22. P. J. Weldon et al., "Benzoquinones from Millipedes Deter Mosquitoes and Elicit Self-Anointing in Capuchin Monkeys (*Cebus* spp.)," *Naturwissenschaften* 90 (2003): 301–304.
23. M. A. Huffman et al., "Leaf-Swallowing by Chimpanzees: A Behavioral Adaptation for the Control of Strongyle Nematode Infections," *International Journal of Primatology* 17 (1996): 475–503.

24. M. Hoogshagen et al., "Mixtures of Milkweed Cardenolides Protect Monarch Butterflies Against Parasites," *Journal of Chemical Ecology* 50 (2024): 52–62.

25. M. S. Singer et al., "Self-Medication as Adaptive Plasticity: Increased Ingestion of Plant Toxins by Parasitized Caterpillars," *PLoS ONE* 4 (2009): e4796.

26. N. F. Milan et al., "Alcohol Consumption as Self-Medication Against Blood-Borne Parasites in the Fruit Fly," *Current Biology* 22 (2012): 488–493.

27. M. L. Phillips et al., "A Specific Neural Substrate for Perceiving Facial Expressions of Disgust," *Nature* 389 (1997): 495–498.

28. V. Curtis and A. Biran, "Dirt, Disgust, and Disease: Is Hygiene in Our Genes?," *Perspectives in Biology and Medicine* 44 (2001): 17–31.

29. S. Stockmaier et al., "An Immune Challenge Reduces Social Grooming in Vampire Bats," *Animal Behavior* 140 (2018): 141–149.

30. Y. Le Conte et al., "Social Immunity in Honeybees (*Apis mellifera*): Transcriptome Analysis of Varroa-Hygienic Behavior," *Insect Molecular Biology* 20 (2011): 399–408.

31. A. Easton et al., "Molecular Evidence of Hybridization Between Pig and Human *Ascaris* Indicates an Interbred Species Complex Infecting Humans," *eLife* 9 (2020): e61562.

32. K. Walton Potter, "How Women Are Bearing the Brunt of the Ebola Epidemic," *Fair Observer*, October 16, 2014, https://www.fairobserver.com/region/africa/how-women-are-bearing-the-brunt-of-the-ebola-epidemic-01745/.

33. R. Wall Kimmerer, "The Serviceberry: An Economy of Abundance," *Emergence Magazine*, October 26, 2022, https://emergencemagazine.org/essay/the-serviceberry/.

6. SYMBIOSIS: THE SECRET EVOLUTIONARY SHORTCUT

1. J. L. Bronstein, "The Evolution of Facilitation and Mutualism," *Journal of Ecology* 98 (2009): 1160–1170.

2. M. Stat et al., "Functional Diversity in Coral-Dinoflagellate Symbiosis," *Proceedings of the National Academy of Sciences USA* 105 (2008): 9256–9261.

3. J. B. Zhao et al., "A Switch from Mutualist to Exploiter Is Reflected in Smaller Egg Loads and Increased Larval Mortalities in a 'Cheater' Fig Wasp," *Acta Oecologica* 57 (2014): 51–57.

4. Y. Ziv and J. L. Bronstein, "Infertile Seeds of *Yucca schottii*: A Beneficial Role for the Plant in the Yucca-Yucca Moth Mutualism?," *Evolutionary Ecology* 10 (1996): 63–76.
5. T. A. Hendry et al., "Ongoing Transposon-Mediated Genome Reduction in the Luminous Bacterial Symbionts of Deep-Sea Ceratioid Anglerfishes," *mBio* 9 (2018): e01033-18.
6. H. Helmby, "Human Helminth Therapy to Treat Inflammatory Disorders: Where Do We Stand?," *BMC Immunology* 16 (2015): 12.
7. S. L. Miller, "A Production of Amino Acids Under Possible Primitive Earth Conditions," *Science* 117 (1953): 528–529.
8. H. J. Bremermann, "Parasites at the Origin of Life," *Journal of Mathematical Biology* 16 (1983): 165–180.
9. R. Mizuuchi and N. Ichihashi, "Primitive Compartmentalization for the Sustainable Replication of Genetic Molecules," *Life* (Basel) 11 (2021): 191.
10. D. Axe, "Error Catastrophe: Manfred Eigen's Show-Stopper Is Still Stopping the Origin-of-Life Show," *Science & Culture Today*, March 19, 2019, https://scienceandculture.com/2019/03/error-catastrophe-manfred-eigens-show-stopper-is-still-stopping-the-origin-of-life-show/.
11. J. P. McCutcheon, "The Bacterial Essence of Tiny Symbiont Genomes," *Current Opinion in Microbiology* 13 (2010): 73–78.
12. B. Heuer, "Mitochondrial DNA: Unraveling the 'Other' Genome," *Journal of the American Association of Nurse Practitioners* 33 (2021): 673–675.
13. N. A. Moran et al., "Regulation of Transcription in a Reduced Bacterial Genome: Nutrient-Provisioning Genes of the Obligates Symbiont *Buchnera aphidicola*," *Journal of Bacteriology* 187 (2005): 4229–4237.
14. M. D. Katinka et al., "Genome Sequence and Gene Compaction of the Eukaryote Parasite *Encephalitozoon cuniculi*," *Nature* 414 (2001): 450–453.
15. E. D. Salomaki et al., "Gregarine Single-Cell Transcriptomics Reveals Differential Mitochondrial Remodeling and Adaptation in Apicomplexans," *BMC Biology* 19 (2021): 77.
16. R. Harada et al., "A Cellular Entity Retaining Only Its Replicative Core: Hidden Archaeal Lineage with an Ultra-Reduced Genome," *bioRxiv*, May 2, 2025, https://doi.org/10.1101/2025.05.02.651781.
17. Y. I. Wolf and E. V. Koonin, "Genome Reduction as the Dominant Mode of Evolution," *Bioessays* 35 (2013): 829–837.
18. J. Muvengwi and E. T. F. Witkowski, "Cascading Effects of Termite Mounds in African Savannas," *New Zealand Journal of Botany* 58 (2020): 167–193.

19. T. Hinzke et al., “Host-Microbe Interactions in the Chemosynthetic *Riftia pachyptila* Symbiosis,” *mBio* 10 (2019): e02243-19.
20. S. Dash et al., “Understanding the Role of the Gut Microbiome in Brain Development and Its Association with Neurodevelopmental Psychiatric Disorders,” *Frontiers in Cell and Developmental Biology* 10 (2022): 880544.
21. A. Kumar et al., “Gut Microbiota in Anxiety and Depression: Unveiling the Relationships and Management Options,” *Pharmaceuticals* (Basel) 16 (2023): 565.
22. P. L. Nguyen and M. van Baalen, “On the Difficult Evolutionary Transition from the Free-Living Lifestyle to Obligate Symbiosis,” *PLoS One* 15 (2020): e0235811.
23. A. A. David, “Climate Change and Shell-Boring Polychaetes (Annelida: Spionidae): Current State of Knowledge and the Need for More Experimental Research,” *Biological Bulletin* 241 (2021): 4–15.
24. While these fungi cannot be grown in pure culture (free of a host plant) under most conditions, they do form spores and hyphae in soil in association with roots. Some analyses suggest that they might germinate spores in the absence of the host but cannot sustain full development, reproduction, or prolonged growth without host-derived carbon and lipids. For more information, see L. Lanfanco and J. P. W. Young, “Genetic and Genomic Glimpses of the Elusive Arbuscular Mycorrhizal Fungi,” *Current Opinion in Plant Biology* 15 (2012): 454–461.
25. A. Olson et al. “Insight Into Trade-Off Between Wood Decay and Parasitism from the Genome of a Fungal Forest Pathogen,” *New Phytologist* 194 (2012): 1001–1013. Note that not all wood-decay fungi come from parasitic ancestors; some may have evolved saprotrophy independently or from free-living saprobes. In addition, many transitions appear to be inferred from phylogeny and comparative genomics, rather than the direct fossil record.
26. L. Tipton et al., “A Developing Symbiosis: Enabling Cross-Talk Between Ecologists and Microbiome Scientists,” *Frontiers in Microbiology* 10 (2019): 292.

7. HITCHHIKERS OF HISTORY

1. World Health Organization, *World Malaria Report 2021*, accessed November 4 2025, https://www.who.int/teams/global-malaria-programme/reports/world-malaria-report-2021.

2. M. Michel et al., "Ancient *Plasmodium* Genomes Shed Light on the History of Human Malaria," *Nature* 631 (2024): 125–233.
3. C. Bern et al., "*Trypanosoma cruzi* and Chagas' Disease in the United States," *Clinical Microbiology Reviews* 24 (2011): 655–681.
4. C. M. Couto-Silva et al., "Indigenous People from Amazon Show Genetic Signatures of Pathogen-Driven Selection," *Science Advances* 9 (2023): eabo0234.
5. M. A. Toups et al., "Origin of Clothing Lice Indicates Early Clothing Use by Anatomically Modern Humans in Africa," *Molecular Biology and Evolution* 28 (2011): 29–32.
6. P. D. Mitchell et al., "Human Intestinal Parasites in Crusader Acre: Evidence for Migration with Disease in the Medieval Period," *International Journal of Paleopathology* 1 (2011): 132–137.
7. J. Piret and G. Boivin, "Pandemics Throughout History," *Frontiers in Microbiology* 11 (2021): 631736.
8. A. Koch et al., "Earth System Impacts of the European Arrival and Great Dying in the Americas After 1942," *Quaternary Science Reviews* 207 (2019): 13–36.
9. R. Barquera et al., "Ancient Genomes Reveal a Deep History of *Treponema pallidum* in the Americas," *Nature* 640 (2024): 186–193.
10. E. Gianchecchi et al., "Yellow Fever: Origin, Epidemiology, Preventative Strategies and Future Prospects," *Vaccines* (Basel) 10 (2022): 372.
11. P. T. Rodrigues et al., "Human Migration and the Spread of Malaria Parasites to the New World," *Scientific Reports* 8 (2018): 1993.
12. J. S. Marr and J. T. Cathey, "The 1802 Saint-Domingue Yellow Fever Epidemic and the Louisiana Purchase," *Journal of Public Health Management and Practice* 19 (2023): 77–82.
13. L. Hurtado et al., "Long-Term Transmission Patterns and Public Health Policies Leading to Malaria Elimination in Panama," *Malaria Journal* 19 (2020): 265.
14. V. J. Cirillo, "Two Faces of Death: Fatalities from Disease and Combat in America's Principal Wars, 1775 to Present," *Perspectives in Biology and Medicine* 51 (2008): 121–133.
15. Official military statistical data from F. H. Mowery, "Statistics of Malaria," chapter 14 in *Infectious Diseases, Volume II* (U.S. Army Medical Department, 1945), AMEDD Center of History & Heritage, https://achh.army.mil/history/book-wwii-infectiousdisvolii-chapter14.

16. S. Paltzer, "The Other Foe: The U.S. Army's Fight Against Malaria in the Pacific Theater, 1942–45," Army Historical Foundation, accessed 4 November, 2025, https://armyhistory.org/the-other-foe-the-u-s-armys-fight-against-malaria-in-the-pacific-theater-1942-45/.
17. J. M. Hayman Jr., "Filariasis," chapter 4 in *Internal Medicine, Volume III* (U.S. Army Medical Department, 1947), AMEDD Center of History & Heritage, https://achh.army.mil/history/book-wwii-internalmedicinevoliii-chapter4.
18. Centers for Disease Control and Prevention, "One Health History," accessed 4 November, 2025. https://www.cdc.gov/one-health/about/one-health-history.html.
19. J. L. Derr, "The Damned Body: Thinking Historically About Water Security and Public Health," *Daedalus* 150 (2021): 143–158.
20. A. C. Hibbs et al., "Irrigation and Infection: The Immunoepidemiology of Schistosomiasis in Ancient Nubia," *American Journal of Physical Anthropology* 145 (2011): 290–298.
21. Acts of the Apostles 12: 21–23 (KJV) states: "And upon a set day Herod, arrayed in royal apparel, sat upon his throne . . . And the people gave a shout, saying It is the voice of a god, and not of a man. And immediately the angel of the Lord smote him, because he gave not God the glory: and he was eaten of worms, and gave up the ghost."
22. R. Espinoza et al., "About the Death of Herod the Great," *Revista médica de Chile 131* (2003): 569–572.
23. D. Langgut, "Mid-7th century BC Human Parasite Remains from Jerusalem," *International Journal of Paleopathology* 36 (2022): 1–6.
24. S. Anderson, "Krimi Roga-Internal Parasitic Disease Through the Lens of Ayurveda," California College of Ayurveda Blog, April 8, 2021, accessed 5 November, 2025, https://www.ayurvedacollege.com/blog/krimi-roga-internal-parasitic-disease-through-the-lens-of-ayurveda/.
25. An excellent source on this subject is M. Harris, *Good to Eat: Riddles of Food and Culture* (Waveland, 1998).
26. B. Bierlich, "Notions and Treatment of Guinea Worm in Northern Ghana," *Social Science & Medicine* 41 (1995): 501–509.
27. U. G. Ekeleme et al., "Nutritional Status and Intestinal Parasitism Determinants of Cognitive Performance Among Secondary School Students in South-East Nigeria," *Clinical Nutrition Open Science* 61 (2025): 200–216.

28. R. P. Parajuli et al., "Risk Factors for *Ascaris lumbricoides* Infection and Its Association with Nutritional Status and IQ in 14-Year-Old Adolescents in Chitwan, Nepal," *Scientific Reports* 14 (2024): 26005.
29. J. B. Madsen, "Barriers to Prosperity: Parasitic and Infectious Diseases, IQ, and Economic Development," *World Development* 78 (2016): 172–187.
30. J. Hackman and D. Hruschka, "Fast Life Histories, Not Pathogens, Account for State-Level Variation in Homicide, Child Maltreatment, and Family Ties in the U.S.," *Evolution and Human Behavior* 34 (2013): 118–124.
31. Proceedings B editorial team, "Retraction: 'Parasite Prevalence and the Worldwide Distribution of Cognitive Ability' (2010), by Eppig, Fincher and Thornhill," *Proceedings of the Royal Society* B 291 (2024): 20240463.
32. E. Miguel and M. Kremer, "Worms: Identifying Impacts on Education and Health in the Busia District, Kenya," *Econometrica* 72, no. 1 (2004): 159–217; S. Baird et al., "Worms at Work: Long-Run Impacts of Child Health Gains," *Quarterly Journal of Economics* 131 (2016): 1637–1680.
33. V. A. Welch et al., "Mass Deworming to Improve Developmental Health and Wellbeing of Children in Low-Income and Middle-Income Countries: A Systematic Review and Network Meta-Analysis," *Lancet Global Health* 5 (2017): E40–E50.
34. C. Elman et al., "Extending Public Health: The Rockefeller Sanitary Commission and Hookworm in the American South," *American Journal of Public Health* 104 (2014): 47–58.
35. P. D. Mitchell, "Human Parasites in the Roman World: Health Consequences of Conquering an Empire," *Parasitology* 144 (2017): 48–58.
36. J. L. Gerberding, "Recent Case of Extensively Drug-Resistant TB: CDC's Public Health Response," CDC Congressional Testimony, June 6, 2007, https://archive.cdc.gov/www_cdc_gov/washington/testimony/2007/t20070606.htm.
37. World Health Organization, "WHO Extends Its SARS-Related Travel Advice to Beijing and Shanxi Province (China) and to Toronto (Canada)," April 23, 2003, https://www.who.int/emergencies/disease-outbreak-news/item/2003_04_23-en.
38. M. Chan, "Ebola Virus Disease in West Africa: No Early End to the Outbreak," *New England Journal of Medicine* 371 (2014): 1183–1185.

39. R. J. Hatchett et al., "Public Health Interventions and Epidemic Intensity During the 1918 Influenza Pandemic," *Proceedings of the National Academy of Sciences of the USA* 104 (2007): 7582–7587.

8. THE GHOSTS OF EVOLUTION PAST

1. M. L. Ledger et al., "Parasite Infection at the Early Farming Community of Catalhoyuk," *Antiquity* 93 (2019): 573–587.
2. R. D. E. MacPhee and P. A. Marx, "Humans, Hyperdisease, and the First-Contact Extinctions," in *Natural Change and Human Impact in Madagascar,* ed. S. M. Goodman and B. D. Patterson (Washington, DC: Smithsonian Institution Press, 1997), 169–217.
3. R. G. Roberts et al., "New Ages for the Last Australian Megafauna: Continent-Wide Extinction About 46,000 Years Ago," *Science* 292 (2001): 1888–1892.
4. C. van Riper III et al., "The Epizootiology and Ecological Significance of Malaria in Hawaiian Land Birds," *Ecological Monographs* 56 (1986): 327–344.
5. S. Altizer et al., "Climate Change and Infectious Diseases: From Evidence to a Predictive Framework," *Science* 341 (2013): 514–519.
6. "By 1750, both (Plasmodium) vivax and falciparum malaria were common from the tropics of Latin America to the Mississippi valley to New England." Institute of Medicine, Committee on the Economics of Antimalarial Drugs, ed. *Saving Lives, Buying Time: Economics of Malaria Drugs in an Age of Resistance* (Washington, DC: National Academies Press, 2004), 126.
7. P. C. Dentzien-Dias et al., "Tapeworm Eggs in a 270-Million-Year-Old Shark Coprolite," *PloS ONE* 8 (2013): e55007.
8. S. L. Sawyer et al., "Positive Selection of Primate *TRIM5*α Identifies a Critical Species-Specific Retroviral Restriction Domain," *Proceedings of the National Academy of Sciences USA* 102 (2005): 2832–2837.
9. L. M. Abegglen et al., "Potential Mechanisms for Cancer Resistance in Elephants and Comparative Cellular Response to DNA Damage in Humans," *JAMA* 314 (2016): 1850–1860.
10. M. Dannemann et al., "Introgression of Neandertal- and Denisovan-like Haplotypes Contributes to Adaptive Variation in Human Toll-like Receptors," *American Journal of Human Genetics* 98 (2016): 22–33.

11. B. C. Scheele et al., "Amphibian Fungal Panzootic Causes Catastrophic Ongoing Loss of Biodiversity," *Science* 363 (2019): 1459–1463.
12. C. D. Harvell et al., "Disease Epidemic and a Marine Heat Wave Are Associated with the Continental-Scale Collapse of a Pivotal Predator (*Pycnopodia helianthoides*)," *Science Advances* 5 (2019): eaau7042.
13. *Nature* 348 (1990): 104.
14. R. R. Dunn et al., "The Sixth Mass Coextinction: Are Most Endangered Species Parasites and Mutualists?," *Proceedings of the Royal Society* B 276 (2009): 3037–3045.
15. C. L. Wood et al., "Marine Protected Areas Facilitate Parasite Populations Among Four Fished Host Species of Central Chile," *Journal of Animal Ecology* 82 (2013): 1276–1287.
16. I. de Buron et al., "ParasiteBlitz: Adaptation of the BioBlitz Concept to Parasitology," *Journal of Helminthology* 99 (2025): 1–14.
17. C. J. Carlson et al., "What Would It Take to Describe the Global Diversity of Parasites?," *Proceedings of the Royal Society* B 287 (2020): 20201841.
18. C. J. Carlson et al., "Parasite Biodiversity Faces Extinction and Redistribution in a Changing Climate," *Science Advances* 3 (2017): e1602422.
19. S. R. Doyle et al., "Population Genomics of Ancient and Modern *Trichuris trichiura*," *Nature Communications* 13 (2022): 3888.
20. A. Araujo et al., "The Role of Mummy Studies in Paleoparasitology," *Chungará* (Arica) 32 (2000): 111–115.
21. G. Fornaciari et al., "Chagas' Disease in Peruvian Inca Mummy," *Lancet* 339 (1992): P128–129.
22. M. Michel et al., "Ancient *Plasmodium* Genomes Shed Light on the History of Human Malaria," *Nature* 631 (2024): 125–133.
23. A. G. Nerlich et al., "Visceral Leishmaniasis During Italian Renaissance, 1522–1562," *Emerging Infectious Diseases* 18 (2012): 184–186.
24. H. Zeberg and S. Paabo, "The Major Genetic Risk Factor for Severe COVID-19 Is Inherited from Neanderthals," *Nature* 587 (2020): 610–612.
25. E. Anastasiou et al., "Prehistoric Schistosomiasis Parasite Found in the Middle East," *Lancet Infectious Diseases* 14 (2014): P553–P554.
26. M. S. Zaghloul et al., "Urinary Schistosomiasis and Associated Bladder Cancer: Update," *Journal of the Egyptian National Cancer Institute* 32 (2020): 44.
27. E. P. K. Parker et al., "Causes of Impaired Oral Vaccine Efficacy in Developing Countries," *Future Microbiology* 13 (2017): 97–118.

28. J. Schölmerich et al., "A Randomised, Double-Blind, Placebo-Controlled Trial of *Trichuris suis* Ova in Active Crohn's Disease," *Journal of Crohn's and Colitis* 11 (2017): 390–399.
29. C. Sgobba, "Would You Eat Parasitic Worm Eggs to Improve Your Health?," *Men's Health*, August 11, 2017, https://www.menshealth.com/health/a19529803/worm-eggs-immune-system/.
30. H. Helmby, "Human Helminth Therapy to Treat Inflammatory Disorders: Where Do We Stand?," *BMC Immunology* 16 (2015): 12.
31. F. Specter, "Mother Reportedly Feeds Teenage Daughter Tapeworms to 'Lose Weight' for Pageants," *Cosmopolitan* (UK), August 22, 2014, https://www.cosmopolitan.com/uk/reports/news/a28996/mother-feeds-tapeworm-beauty-pageant-teenage-daughter/.
32. A. M. Kuris et al., "Ecosystem Energetic Implications of Parasite and Free-Living Biomass in Three Estuaries," *Nature* 454 (2008): 515–518.
33. R. Claerhoudt, "The Legal Protection of Animal Parasites in International Biodiversity Law," *Journal of Environmental Law* 36 (2024): 301–322.
34. C. J. Carlson et al., "A Global Parasite Conservation Plan," *Biological Conservation* 250 (2020): 108596.
35. A. J. Lymbery and N. J. Smith, "Conservation of Parasites: A Primer," *International Journal for Parasitology: Parasites and Wildlife* 21 (2023): 255–263.

9. THE FUTURE OF EVOLUTION'S PUPPETEERS

1. P. A. Leighton et al., "Predicting the Speed of Tick Invasion: An Empirical Model of Range Expansion for the Lyme Disease Vector *Ixodes scapularis* in Canada," *Journal of Applied Ecology* 49 (2012): 457–464.
2. T. Cook et al., "The Relationship Between Increasing Sea-Surface Temperature and the Northward Spread of *Perkinsus marinus* (Dermo)," *Estuarine, Coastal and Shelf Science* 46 (1998): 587–597.
3. M. Titus et al., "First Molecular Characterization of *Proctoeces maculatus* (Looss, 1901) (Digenea: Fellodistomidae) Infecting Blue Mussels (*Mytilus edulis*) in the Northeastern USA," *Journal of Helminthology* 99 (2025): e25.
4. United Nations, Department of Economic and Social Affairs. "68% of World Population Projected to Live in Urban Areas by 2050, Says UN," 2018, https://www.un.org/uk/desa/68-world-population-projected-live-urban-areas-2050-says-un.

5. One Health High-Level Expert Panel (OHHLEP) et al., "One Health: A New Definition for a Sustainable and Healthy Future," *PLoS Pathogens* 18 (2022): e1010537.
6. Y. Ni et al., "A Target-Based Discovery from a Parasitic Helminth as a Novel Therapeutic Approach for Autoimmune Diseases," *eBioMedicine* 95 (2023): 104751.
7. A. Banerjee et al. "Live Attenuated *Leishmania donovani* Centrin Gene-Deleted Parasites Induce IL-23-Dependent IL-17-Protective Immune Response Against Visceral Leishmaniasis in a Murine Model," *Journal of Immunology* 200 (2019): 163–176.
8. O. A. C. Lamers et al., "The Path from Early– to Late–Liver Stage Arresting Genetically Attenuated Parasites as a Malaria Vaccination Strategy," *NPJ Vaccines* 10 (2025): 218.
9. K. M. Esvelt et al., "Concerning RNA-Guided Gene Drives for the Alteration of Wild Populations," *Elife* 3 (2014): e03401.
10. UPI, "American Student Tried to Kill Roommates with Pig Parasite," *Griffin* (Georgia) *Daily News*, February 27, 1970, https://gahistoricnewspapers.galileo.usg.edu/lccn/sn83009936/1970-02-27/ed-1/seq-1/.
11. E. H. Soice et al., "Can Large Language Models Democratize Access to Dual-Use Biotechnology," *arXiv* (2023): 2306.03809.
12. T. Kulken, "Artificial Intelligence in the Biological Sciences: Uses, Safety, Security, and Oversight," CRS Report R47849, November 22, 2023, https://www.congress.gov/crs-product/R47849.
13. The Global Fund, "Gates Foundation," accessed May 11, 2026, https://www.theglobalfund.org/en/private-sector-and-philanthropy/profiles/gates-foundation/.
14. R. M. Holdo et al., "A Disease-Mediated Trophic Cascade in the Serengeti and Its Implications for Ecosystem C," *PLoS Biology* 7 (2009): e1000210.

BIBLIOGRAPHY

Akinyi, M. Y., J. Tung, M. Jeneby, N. B. Patel, J. Altmann, and S. C. Alberts. "Role of Grooming in Reducing Tick Load in Wild Baboons (*Papio cynocephalus*)." *Animal Behaviour* 85 (2013): 559–568.

Angum, F., T. Khan, J. Kaler, L. Siddiqui, and A. Hussain. "The Prevalence of Autoimmune Disorders in Women: A Narrative Review." *Cureus* 12 (2020): e8094.

Aronson, N., B. L. Herwaldt, M. Libman, et al. "Diagnosis and Treatment of *Leishmaniasis*: Clinical Practice Guidelines by the Infectious Diseases Society of America (IDSA) and the American Society of Tropical Medicine and Hygiene (ASTMH)." *Clinical Infectious Diseases* 63 (2016): 1539–1557.

Bell, K., and S. R. Bordenstein. "A Margulian View of Symbiosis and Speciation: The Nasonia Wasp System." *Symbiosis* 87 (2022): 3–10.

Berdoy, M., J. P. Webster, and D. W. Macdonald. "Fatal Attraction in Rats Infected with *Toxoplasma gondii*." *Proceedings of the Royal Society of London B* 267 (2000): 1591–1594.

Biron, D. G., L. Marche, F. Ponton, et al. "Behavioral Manipulation in a Grasshopper Harbouring Hairworm: A Proteomics Approach." *Proceedings of the Royal Society B* 272 (2005): 2117–2126.

Black, J. A., J. L. Reis-Cunha, A. K. Cruz, and L. R. O. Tosi. "Life in Plastic, It's Fantastic! How Leishmania Exploit Genome Instability to Shape Gene Expression." *Frontiers in Cellular and Infection Microbiology* 13 (2023): 1102462.

Boillat, M., P.-M. Hammoudi, S. K. Dogga, et al. "Neuroinflammation-Associated Aspecific Manipulation of Mouse Predator Fear by *Toxoplasma gondii*." *Cell Reports* 30 (2020): 320–334.

Bollani, L., C. Auriti, C. Achille, et al. "Congenital Toxoplasmosis: The State of the Art." *Frontiers in Pediatrics* 10 (2022): 894573.

Bouznach, A., N. Edery, E. Kelmer, N. Shicaht, T. Waner, and S. Perl. "Systemic *Toxoplasma gondii* Infection in a Cat with Incidental Cholangioma." *Israel Journal of Veterinary Medicine* 70 (2015): 64–67.

Bremermann, H. J. "Parasites at the Origin of Life." *Journal of Mathematical Biology* 16 (1983): 165–180.

Bronstein, J. L. "The Evolution of Facilitation and Mutualism." *Journal of Ecology* 98 (2009): 1160–1170.

Brown, G. D. A., L. Walasek, T. L. Mullet, et al. "Political Attitudes and Disease Threat: Regional Pathogen Stress Is Associated with Conservative Ideology Only for Older Individuals." *Personality and Social Psychology Bulletin* 50 (2024): 1581–1595.

Calero-Bernal, R., and S. M. Gennari. "Clinical Toxoplasmosis in Dogs and Cats: An Update." *Frontiers in Veterinary Science* 6 (2019): 54.

Card, D. C., B. W. Perry, R. H. Adams, et al. "Novel Ecological and Climatic Conditions Drive Rapid Adaptation in Invasive Florida Burmese Pythons." *Molecular Ecology* 27 (2018): 4744–4757.

Claessens, A., W. L. Hamilton, M. Kekre, et al. "Generation of Antigenic Diversity in *Plasmodium falciparum* by Structured Rearrangement of Var Genes During Mitosis." *PLoS Genetics* 10 (2014): e1004812.

Clementi, A., G. Battaglia, M. Floris, P. Castellino, C. Ronco, and D. N. Cruz. "Renal Involvement in *Leishmaniasis*: A Review of the Literature." *NDT Plus* 4 (2011): 147–152.

Currie, T. E., and R. Mace. "Analyses Do Not Support the Parasite-Stress Theory of Human Sociality." *Behavioral and Brain Sciences* 35 (2012): 83–85.

Curtis, V., and A. Biran. "Dirt, Disgust, and Disease: Is Hygiene in Our Genes?" *Perspectives in Biology and Medicine* 44 (2001): 17–31.

David, A. A. "A Student-Centered Framework for Teaching Undergraduate Parasitology." *Trends in Parasitology* 33 (2017): P420–P423.

Dawkins, R. "Extended Phenotype—but Not Too Extended: A Reply to Laland, Turner and Jablonka." *Biology and Philosophy* 19 (2004): 377–396.

De Bekker, C., I. Will, B. Das, and R. M. M. Adams. "The Ants (*Hymenoptera: Formicidae*) and Their Parasites: Effects of Parasitic Manipulations and Host Responses on Ant Behavioral Ecology." *Myrmecological News* 28 (2018): 1–24.

de Koning, A. P. J., W. Gu, T. A. Castoe, M. A. Batzer, and D. D. Pollock. "Repetitive Elements May Comprise Over Two-Thirds of the Human Genome." *PLoS Genetics* 7 (2011): e1002384.

Deen, J., M. A. Mengel, and J. D. Clemens. "Epidemiology of Cholera." *Vaccine* 38 (2020): A31–A40.

Dobson, A. P., K. Lafferty, A. M. Kuris, R. F. Hechinger, and W. Jetz. "Homage to Linnaeus: How Many Parasites? How Many Hosts?" *Proceedings of the National Academy of Sciences of the USA* 105 (2008): 11482–11489.

Dorlo, T. P. C., M. Balasegaram, J. H. Beijnen, and P. J. de Vries. "Miltefosine: A Review of Its Pharmacology and Therapeutic Efficacy in the Treatment of Leishmaniasis." *Journal of Antimicrobial Chemotherapy* 67 (2012): 2576–2579.

Douanne, N., G. Dong, A. Amin, et al. "*Leishmania* Parasites Exchange Drug-Resistance Genes Through Extracellular Vesicles." *Cell Reports* 40 (2022): 111121.

Dunne, J. A., K. D. Lafferty, A. P. Dobson, et al. "Parasites Affect Food Web Structure Primarily Through Increased Diversity and Complexity." *PLoS Biology* 11 (2013): e1001579.

Easton, A., S. Gao, S. P. Lawton, et al. "Molecular Evidence of Hybridization Between Pig and Human *Ascaris* Indicates an Interbred Species Complex Infecting Human." *eLife* 9 (2020): e61562.

Fincher, C. L., and R. Thornhill. "Parasite-Stress Promotes In-Group Assortative Sociality: The Cases of Strong Family Ties and Heightened Religiosity." *Behavioral and Brain Sciences* 35 (2012): 61–79.

Fincher, C. L., R. Thornhill, D. R. Murray, and M. Schaller. "Pathogen Prevalence Predicts Human Cross-Cultural Variability in Individualism/Collectivism." *Proceedings of the Royal Society B* 275 (2008): 1279–1285.

Garushyants, S. K., I. B. Rogozin, and E. V. Koonin. "Template Switching and Duplications in SARS-Cov-2 Genomes Give Rise to Insertion Variants That Merit Monitoring." *Communications Biology* 4 (2021): 1343.

Green, D. P. L. "Infectious Diseases, Imposing Density-Dependent Mortality on MHC/HLA Variation, Can Account for Balancing Selection and MHC/HLA Polymorphism." arXiv:2501.00767 (2025).

Hamilton, W. D. "Sex Versus Non-Sex Versus Parasite." *Oikos* 35 (1980): 282–290.

Hammoud, C., A. Kayenbergh, J. Tumusiime, et al. "Trematode Infections Affects Shell Shape and Size in *Bulinus tropicus*." *International Journal for Parasitology: Parasites and Wildlife* 18 (2022): 300–311.

Harada, R., Y. Nishimura, M. Nomura, et al. "A Cellular Entity Retaining Only Its Replicative Core: Hidden Archaeal Lineage with an Ultra-Reduced Genome." *bioRxiv* (2025). https://doi.org/10.1101/2025.05.02.651781.

Harnett, W., and M. M. Harnett. "Epigenetic Changes Induced by Parasitic Worms and Their Excretory-Secretory Products." *Biochemical Society Transactions* 52 (2024): 55–63.

Hartfield, M., and P. D. Keightley. "Current Hypotheses for the Evolution of Sex and Recombination." *Integrative Zoology* 7 (2012): 192–209.

Hayward, A., C. K. Cornwallis, and P. Jern. "Pan-Vertebrate Comparative Genomics Unmasks Retrovirus Macroevolution." *Proceedings of the National Academy of Sciences of the USA* 112 (2015): 464–469.

Hechinger, R. F. "Parasites Help Find Universal Ecological Roles." *Proceedings of the National Academy of Sciences of the USA* 112 (2015): 1656–1657.

Helgen, J. Peril in the Ponds: Deformed Frogs, Politics, and a Biologist's Quest. University of Massachusetts Press, 2012.

Helmby, H. "Human Helminth Therapy to Treat Inflammatory Disorders—Where Do We Stand?" *BMC Immunology* 16 (2015): 12.

Hendry, T. A., L. L. Freed, D. Fader, D. Fenolio, T. T. Sutton, and J. V. Lopez. "Ongoing Transposon-Mediated Genome Reduction in the Luminous Bacterial Symbionts of Deep-Sea Ceratioid Anglerfishes." *mBio* 9 (2018): e01033–18.

Herniou, E. A., E. Huguet, J. Theze, A. Bezier, G. Periquet, and J.-M. Drezen. "When Parasitic Wasps Hijacked Viruses: Genomic and Functional Evolution of Polydnaviruses." *Philosophical Transactions of the Royal Society B* 368 (2013): 20130051.

Heuer, B. "Mitochondrial DNA: Unraveling the 'Other' Genome." *Journal of the American Association of Nurse Practitioners* 33 (2021): 673–675.

Hoogshagen, M., A. P. Hastings, J. Chavez, et al. "Mixtures of Milkweed Cardenolides Protect Monarch Butterflies Against Parasites." *Journal of Chemical Ecology* 50 (2024): 52–62.

Hooshyar, H., and P. Rostamkhani. "Accurate Laboratory Diagnosis of Human Intestinal and Extra-Intestinal Amoebiasis." *Gastroenterology and Hepatology from Bed to Bench* 15 (2022): 343–359.

Huffman, M. A., J. E. Page, M. V. K. Sukhdeo, et al. "Leaf-Swallowing by Chimpanzees: A Behavioral Adaptation for the Control Of Strongyle Nematode Infections." *International Journal of Primatology* 17 (1996): 475–503.

Hurst, G. D. D., and C. L. Frost. "Reproductive Parasitism: Maternally Inherited Symbionts in a Biparental World." *Cold Spring Harbor Perspectives in Biology* 7 (2015): a017699.

Jedryczka, W. "Can History of Parasitic Diseases Increase Social Conservatism? Testing Behavioral Immune System Theory." *Journal of Education Culture and Society* 13 (2022): 383–394.

Johnson, P. T. J., K. B. Lunde, E. G. Ritchie, and A. E. Launer. "The Effect of Trematode Infection on Amphibian Limb Development and Survivorship." *Science* 284 (1999): 802–804.

Jonason, P. K., and D. P. Schmitt. "Quantifying Common Criticisms of Evolutionary Psychology." *Evolutionary Psychological Science* 2 (2016): 177–188.

Kajimoto, A., A. Iwasaki, T. Ohira, and K. Toyota. "Morphological Feminization in Hermit Crabs (Family Paguridae) Induced by Rhizocephalan Barnacles." *Zoological Letters* 11 (2025): 6.

Katinka, M. D., S. Duprat, E. Cornillot, et al. "Genome Sequence and Gene Compaction of the Eukaryote Parasite *Encephalitozoon cuniculi*." *Nature* 414 (2001): 450–453.

Kelly, C. D., A.M. Stoehr, C. Nunn, K. N. Smyth, and Z. M. Prokop. "Sexual Dimorphism in Immunity Across Animals: A Meta-Analysis." *Ecology Letters* 21 (2018): 1885–1894.

Kent, A. G., A. C. Vill, Q. Shi, M. J. Satlin, and I. L. Brito. "Widespread Transfer of Mobile Antibiotic Resistance Genes Within Individual Gut Microbiomes Revealed Through Bacterial Hi-C." *Nature Communications* 11 (2020): 4379.

Klunk, J., T. P. Vilgalys, C. E. Demeure, et al. "Evolution of Immune Genes Is Associated with the Black Death." *Nature* 611 (2022): 312–319.

Kopecky, R., L. Priplatova, S. Boschetti, K. Talmont-Kaminski, and J. Flegr. "Le Petit Machiavellian Prince: Effects of Latent Toxoplasmosis on Political Beliefs and Values." *Evolutionary Psychology* 20 (2022): 14747049221112657.

Kuroda, K., T. Kuroda, H. Nishino, and Y. Takami. "Sex-specific Manipulation of Sexually Cannibalistic Mantid Mating Behavior by Hairworms." *Behavioral Ecology* 35 (2024): arae071.

Lafferty, K. D. "Can the Common Brain Parasite, *Toxoplasma gondii*, Influence Human Culture." *Proceedings of the Royal Society B* 273 (2006): 2749–2755.

Lafferty, K. D., A. P. Dobson, and A. M. Kuris. "Parasites Dominate Food Web Links." *Ecology* 103 (2006): 11211–11216.

Lavialle, C., G. Cornelis, A. Dupressoir, et al. "Paleovirology of 'Syncytins,' Retroviral Env Genes Exapted for a Role in Placentation." *Philosophical Transactions of the Royal Society B* 368 (2013): 20120507.

Le Conte, Y., C. Alaux, J.-F. Martin, et al. "Social Immunity in Honeybees (*Apis mellifera*): Transcriptome Analysis of Varroa-Hygienic Behavior." *Insect Molecular Biology* 20 (2011): 399–408.

Lenski, R. E., M. R. Rose, S. C. Simpson, and S. C. Tadler. "Long-Term Experimental Evolution in *Escherichia coli*: I. Adaptation and Divergence During 2,000 Generations." *American Naturalist* 138 (1991): 1315–1341.

Lindsay, J. A. "*Staphylococcus aureus* Genomics and the Impact of Horizontal Gene Transfer." *International Journal of Medical Microbiology* 304 (2014): 103–109.

Loiseau, C., R. Zoorob, A. Robert, O. Chastel, R. Julliard, and G. Sorci. "*Plasmodium relictum* Infection and MHC Diversity in the House Sparrow (*Passer domesticus*)." *Proceedings of the Royal Society B* 278 (2011): 1264–1272.

Luzzatto, L., M. Ally, and R. Notaro. "Glucose-6-phosphate Dehydrogenase Deficiency." *Blood* 136 (2020): 1225–1240.

Mao, J., Q. Zhang, and Y.-S. Cong. "Human Endogenous Retroviruses in Development and Disease." *Computational and Structural Biotechnology Journal* 19 (2021): 5978–5986.

McCutcheon, J. P. "The Bacterial Essence of Tiny Symbiont Genomes." *Current Opinion in Microbiology* 13 (2010): 73–78.

McDonald, R. A., J. K. Wilson-Aggarwal, G. J. F. Swan, et al. "Ecology of Domestic Dogs *Canis familiaris* as an Emerging Reservoir of Guinea Worm *Dracunculus medinensis* Infection." *PLoS Neglected Tropical Diseases* 14 (2020): e0008170.

McLaren, P. J., and J. Fellay. "HIV-1 and Human Genetic Variation." *Nature Reviews Genetics* 22 (2021): 645–657.

McMillan, L. E. M., R. H. Herbison, D. G. Biron, et al. "The Caterpillar *Manduca sexta* Brain Shows Changes in Gene Expression and Protein Abundance Correlating with Parasitic Manipulation of Behavior." *Scientific Reports* 14 (2024): 31773.

Mewamba, E. M., O. A. Nyangiri, H. A. Noyes, M. Egesa, E. Matovu, and G. Simo. "The Genetics of Human Schistosomiasis Infection Intensity and Liver Disease: A Review." *Frontiers in Immunology* 12 (2021): 612468.

Milan, N. F., B. Z. Kacsoh, and T. A. Schlenke. "Alcohol Consumption as Self-Medication Against Blood-Borne Parasites in the Fruit Fly." *Current Biology* 22 (2012): 488–493.

Milinski, M. "The Function of Mate Choice in Sticklebacks: Optimizing MHC Genes." *Journal of Fish Biology* 63 (2003): 1–16.

Milinski, M., and T. C. M. Bakker. "Female Sticklebacks Use Male Coloration in Mate Choice and Hence Avoid Parasitized Males." *Nature* 344 (1990): 330–333.

Miller, M. A., J. M. Kinsella, R. W. Snow, et al. "Parasite Spillover: Indirect Effects of Invasive Burmese Pythons." *Ecology and Evolution* 8 (2017): 830–840.

Miller, S. L. "A Production of Amino Acids Under Possible Primitive Earth Conditions." *Science* 117 (1953): 528–529.

Mishina, T., M.-C. Chui, Y. Hashiguchi, et al. "Massive Horizontal Gene Transfer and the Evolution of Nematomorph-Driven Behavioral Manipulation of Mantids." *Current Biology* 33 (2023): 4988–4994.

Mizuuchi, R., and N. Ichihashi. "Primitive Compartmentalization for the Sustainable Replication of Genetic Molecules." *Life* (Basel) 11 (2021): 191.

Moran, N. A., H. E. Dunbar, and J. L. Wilcox. "Regulation of Transcription in a Reduced Bacterial Genome: Nutrient-Provisioning Genes of the Obligate Symbiont *Buchnera aphidicola*." *Journal of Bacteriology* 187 (2005): 4229–4237.

Muehlenbachs, A., J. Bhatnagar, C. A. Agudelo, et al. "Malignant Transformation of *Hymenolepis nana* in a Human Host." *New England Journal of Medicine* 373 (2015): 1845–1852.

Murray, D. R., M. Schaller, and P. Suedfeld. "Pathogens and Politics: Further Evidence That Parasite Prevalence Predicts Authoritarianism." *PLoS ONE* 8 (2013): e62275.

Nasir, A., and G. Caetano-Anolles. "A Phylogenomic Data-Driven Exploration of Viral Origins and Evolution." *Science Advances* 1 (2015): e1500527.

Nguyen, A. V., and A. M. Soulika. "The Dynamics of the Skin's Immune System." *International Journal of Molecular Sciences* 20 (2019): 1811.

Pagel, M., and W. Bodmer. "A Naked Ape Would Have Fewer Parasites." *Proceedings of the Royal Society of London* B 270 (2003): S117–S119.

Petsko, G. A. "Transformation." *Genome Biology* 7 (2006): 117.

Phillips, M. L., A. W. Young, C. Senior, et al. "A Specific Neural Substrate for Perceiving Facial Expressions of Disgust." *Nature* 389 (1997): 495–498.

Poirotte, C., F. Massol, A. Herbert, et al. E. "Mandrills Use Olfaction to Socially Avoid Parasitized Conspecifics." *Science Advances* 3 (2017): e1601721.

Pozio, E. "Adaptation of *Trichinella spp.* for Survival in Cold Climates." *Food and Waterborne Parasitology* 4 (2016): 4–12.

Quach, H., D. Wilson, G. Laval, et al. "Different Selective Pressures Shape the Evolution of Toll-Like Receptors in Human and African Great Ape Populations." *Human Molecular Genetics* 22 (2013): 4829–4840.

Radwan, J., W. Babik, J. Kaufman, T. L. Lenz, and J. Winternitz. "Advances in the Evolutionary Understanding of MHC Polymorphism." *Trends in Genetics* 36 (2020): 298–311.

Recker, M., C. O. Buckee, A. Serazin, et al. "Antigenic Variation in Plasmodium Falciparum Malaria Involves a Highly Structured Switching Pattern." *PLoS Pathogens* 7 (2011): e1001306.

Rolff, J. "Why Did the Acquired Immune System of Vertebrates Evolve?" *Developmental & Comparative Immunology* 31 (2007): 476–482.

Sallard, E., J. Halloy, D. Casane, E. Decroly, and J. van Helden. "Tracing the Origins of SARS-CoV-2 in Coronavirus Phylogenies: A Review." *Environmental Chemistry Letters* 19 (2021): 769–785.

Salomaki, E. D., K. X. Terpis, S. Rueckert, et al. "Gregarine Single-Cell Transcriptomics Reveals Differential Mitochondrial Remodeling and Adaptation in Apicomplexans." *BMC Biology* 19 (2021): 77.

Sanchez, S. G., and S. Besteiro. "The Pathogenicity and Virulence of *Toxoplasma gondii*." *Virulence* 12 (2021): 3095–3114.

Santin, A. D., D. E. Scheim, P. A. McCullough, M. Yagisawa, and T. J. Borody. "Ivermectin: A Multifaceted Drug of Nobel Prize–Honoured Distinction with Indicated Efficacy Against a New Global Scourge, COVID-19." *New Microbes and New Infections* 43 (2021): 100924.

Scherf, A., J. J. Lopez-Rubio, and L. Riviere. "Antigenic Variation in *Plasmodium falciparum*." *Annual Review of Microbiology* 62 (2008): 445–470.

Sharun, K., K. Dhama, A. M. Pawde, et al. "SARS-CoV-2 in Animals: Potential for Unknown Reservoir Hosts and Public Health Implications." *Veterinary Quarterly* 41 (2021): 181–201.

Shetty, A., S. Shetty, and O. Dsouza. "Medical Symbols in Practice: Myths Vs. Reality." *Journal of Clinical and Diagnostic Research* 9 (2014): PC12–PC14.

Simonetti, O., V. Zerbato, S. Di Bella, R. Luzzati, and F. Cavalli. "*Dracunculiasis* Over the Centuries: The History of a Parasite Unfamiliar to the West." *Le Infezioni in Medicina* 31 (2023): 257–264.

Singer, M. S., K. C. Mace, and E. A. Bernays. "Self-Medication as Adaptive Plasticity: Increased Ingestion of Plant Toxins by Parasitized Caterpillars." *PLoS ONE* 4 (2009): e4796.

Soucy, S. M., J. Huang, and P. Gogarten. "Horizontal Gene Transfer: Building the Web of Life." *Nature Reviews Genetics* 16 (2015): 472–482.

Srivastava, S., J. Mishra, A. K. Gupta, A. Singh, P. Shankar, and S. Singh. "Laboratory Confirmed Miltefosine Resistant Cases of Visceral *Leishmaniasis*." *Parasites & Vectors* 10 (2017): 49.

Stat, M., E. Morris, and R. D. Gates. "Functional Diversity in Coral-Dinoflagellate Symbiosis." *Proceedings of the National Academy of Sciences of the USA* 105 (2008): 9256–9261.

Steverding, D. "The History of *Leishmaniasis*." *Parasites & Vectors* 10 (2017): 82.

Stiemsma, L. T., L. A. Reynolds, S. E. Turvey, and B. B. Finlay. "The Hygiene Hypothesis: Current Perspectives and Future Therapies." *ImmunoTargets and Therapy* (2015): 143–157.

Stockmaier, S., D. I. Bolnick, R. A. Page, and G. G. Carter. "An Immune Challenge Reduces Social Grooming in Vampire Bats." *Animal Behaviour* 140 (2018): 141–149.

Strube, C., K. Raue, and E. Janecek. "Simple, but Not Easy—Opportunities and Challenges From Teachers' and Students' Perspectives in the 21st Century of Veterinary Parasitology Teaching." *Veterinary Parasitology* 252 (2018): 74–79.

Sun, S., B. La Scola, V. D. Bowman, et al. "Structural Studies of the Sputnik Virophage." *Journal of Virology* 84 (2010): 894–897.

Tong, W. H., C. Pavey, R. O'Handley, and A. Vyas. "Behavioral Biology of *Toxoplasma gondii* Infection." *Parasites & Vectors* 14 (2021): 77.

Torgerson, P. R., and P. Mastroiacovo. "The Global Burden of Congenital Toxoplasmosis: A Systematic Review." *Bulletin of the World Health Organization* 91 (2013): 501–508.

Uliana, S. R. B., J. C. Ruiz, and A. K. Cruz. "Leishmania Genomics: Where Do We Stand?" In *Bioinformatics in Tropical Disease Research: A Practical and Case-Study Approach*, ed. A. Gruber, A. M. Durham, C. Huynh, et al., chapter B02. National Center for Biotechnology Information, 2008.

Van Valen, L. "A New Evolutionary Law." *Evolutionary Theory* 1 (1973): 1–30.

Vergara, D., C. M. Lively, K. C. King, and J. Jokela. "The Geographic Mosaic of Sex and Infection in Lake Populations of a New Zealand Snail at Multiple Spatial Scales." *American Naturalist* 182 (2013): 484–492.

Vikeved, E., A. Backlund, and C. Alsmark. "The Dynamics of Lateral Gene Transfer in the Genus Leishmania—A Route for Adaptation and Species Diversification." *PLoS Neglected Tropical Diseases* 10 (2016): e0004326.

Waldor, M. K., and J. J. Mekalanos. "Lysogenic Conversion by a Filamentous Phage Encoding Cholera Toxin." *Science* 272 (1996): 1910–1914.

Wang, J., X. Lu, W. Zhang, and G.-H. Liu. "Endogenous Retroviruses in Development and Health." *Trends in Microbiology* 32 (2024): 342–354.

Wang, Z. T., S. K. Verma, J. P. Dubey, and L. D. Sibley. "The Aromatic Amino Acid Hydroxylase Genes AAH1 and AAH2 in *Toxoplasma gondii* Contribute to Transmission in the Cat." *PLoS Pathogens* 13 (2017): e1006272.

Warren, K. S., and A. A. F. Mahmoud. "Tapeworms." In *Geographic Medicine for the Practitioner*, 2nd ed., 182–189. Springer, 1985.

Watson, J. D., and F. H. C. Crick. "Molecular Structure of Nucleic Acids: A Structure for Deoxyribose Nucleic Acid." *Nature* 171 (1953): 737–738.

Wegner, K. M., M. Kalbe, M. Milinski, and T. B. H. Reusch. "Mortality Selection During the 2003 European Heat Wave in Three-Spined Sticklebacks: Effects of Parasites and MHC Genotype." *BMC Evolutionary Biology* 8 (2008): 124.

Weinstein, S. B., and A. M. Kuris. "Independent Origins of Parasitism in Animalia." *Biology Letters* 12 (2016): 20160324.

Weldon, P. J., J. R. Aldrich, J. A. Klun, J. E. Oliver, and M. Debboun. "Benzoquinones from Millipedes Deter Mosquitoes and Elicit Self-Anointing in Capuchin Monkeys (*Cebus spp.*)." *Naturwissenschaften* 90 (2003): 301–304.

Wohlfert, E. A., I. J. Blader, and E. H. Wilson. "Brains and Brawn: Toxoplasma Infections of the Central Nervous System and Skeletal Muscle." *Trends in Parasitology* 33 (2017): 519–531.

Xiao, J., G. Kannan, L. Jones-Brando, et al. "Sex-Specific Changes in Gene Expression and Behavior Induced by Chronic Toxoplasma Infection in Mice." *Neuroscience* 206 (2012): 39–48.

Xu, F., J. Jerlstrom-Hultqvist, M. Kolisko, et al. "On the Reversibility of Parasitism: Adaptation to a Free-Living Lifestyle Via Gene Acquisitions in the Diplomonad *Trepomaonas* sp. PC1." *BMC Biology* 14, 62 (2016).

Zarlenga, D., P. Thompson, M. Mitreva, B. A. Rosa, and E. Hoberg. "Horizontal Gene Transfer Provides Insights Into the Deep Evolutionary History and Biology of *Trichinella*." *Food and Waterborne Parasitology* 27 (2022): e00155.

Zhan, Z., Q. He, K. E. Tiedje, K. P. Day, and M. Pascual. "Hyper-Diverse Antigenic Variation and Resilience to Transmission-Reducing Intervention in Falciparum Malaria." *Nature Communications* 15 (2024): 7343.

Zhao, J.-B., Y.-Q. Peng, R. J. Quinnell, S. G. Compton, and D.-R. Yang. "A Switch from Mutualist to Exploiter Is Reflected in Smaller Egg Loads and

Increased Larval Mortalities in a 'Cheater' Fig Wasp." *Acta Oecologica* 57 (2014): 51–57.

Ziv, Y., and J. L. Bronstein. "Infertile Seeds of *Yucca schottii*: A Beneficial Role for the Plant in the Yucca–Yucca Moth Mutualism?" *Evolutionary Ecology* 10 (1996): 63–76.

Zou, Y., L. Pu, A. Guo, et al. "Helminth Reshapes Host Gut Microbiota and Immunoregulation by Deploying an Antimicrobial Program of Innate Immunity." *Gut Microbes* 17 (2025): 2496447.

INDEX